T0137201

Intelligent Systems Reference Library

Volume 105

Series editors

Janusz Kacprzyk, Polish Academy of Sciences, Warsaw, Poland
e-mail: kacprzyk@ibspan.waw.pl

Lakhmi C. Jain, Bournemouth University, Fern Barrow, Poole, UK, and
University of Canberra, Canberra, Australia
e-mail: jainlc2002@yahoo.co.uk

Anna Esposito · Lakhmi C. Jain
Editors

Toward Robotic Socially Believable Behaving Systems - Volume I

Modeling Emotions

Editors
Anna Esposito
Department of Psychology
Seconda Università di Napoli and IIASS
Caserta
Italy

Lakhmi C. Jain
Faculty of Science and Technology
Data Science Institute, Bournemouth
  University
Fern Barrow, Poole
UK

ISSN 1868-4394   ISSN 1868-4408   (electronic)
Intelligent Systems Reference Library
ISBN 978-3-319-80951-9   ISBN 978-3-319-31056-5   (eBook)
DOI 10.1007/978-3-319-31056-5

This Springer imprint is published by Springer Nature
The registered company is Springer International Publishing AG Switzerland

Preface

This volume is a collection of research studies on the attempt to model emotions in complex autonomous systems. Several experts in the field are reporting their efforts and reviewing the literature in order to shed light on how the processes of coding and decoding emotional states took place in humans, which are the physiological, physical, and psychological variables involved, invent new mathematical models and algorithms to describe them, and motivate these investigations at the light of observable societal changes and needs, such as the aging population and the cost of health care services. The consequences are the implementation of emotionally and socially believable machines, acting as helpers into domestic spheres, where emotions drive behaviors and actions.

The implementation of such complex autonomous systems covers a wide collection of research problems and encompasses a wide range of social, theoretical and technological outcomes. Such heterogeneous research, requiring expertise from several, yet complementary scientific domains (such as cognitive, computational, and information communication technology sciences) can be developed only within a flexible and ample research coordination that pointed out what is needed from each complementary domain and which challenges the research should face in order to produce breakthroughs, cross-fertilization, and advances within constituent disciplines. We believe that the content of this volume will contribute to such an effort.

The editors would like to thank the contributors and the International Scientific Committee of reviewers listed below for their rigorous and invaluable scientific revisions, dedication, and priceless selection process. Thanks are also due to the Springer-Verlag for their excellent support during the development phase of this research book.

Italy
Australia

Anna Esposito
Lakhmi C. Jain

International Scientific Committee

- Samantha Adams (Plymouth University, UK)
- Samer Al Moubayed (School of Computer Science and Communication, Stockholm, Sweden)
- Ivana Baldassarre (Seconda Università di Napoli, Italy)
- Tony Belpaeme (Plymouth University, UK)
- Štefan Beňuš (Constantine the Philosopher University, Nitra, Slovakia)
- Ronald Böck (Otto von Guericke University Magdeburg, Germany)
- Branislav Borova (University of Novi Sad, Serbia)
- Nikolaos Bourbakis (Wright State University, Dayton, USA)
- Angelo Cafaro (Telecom ParisTech, Paris, France)
- Angelo Cangelosi (Plymouth University, UK)
- Chloe Clavel (Telecom ParisTech, Paris, France)
- Gennaro Cordasco (Seconda Università di Napoli and IIASS, Italy)
- Conceição Cunha (Ludwig-Maximilians University of Munich, Germany)
- Alessandro Di Nuovo (Plymouth University, UK)
- Thomas Drugman (University of Mons, Belgium)
- Stéphane Dupont (University of Mons, Belgium)
- Anna Esposito (Seconda Università di Napoli and IIASS, Italy)
- Antonietta Maria Esposito (Osservatorio Vesuviano, Napoli, Italy)
- Marcos Faúndez-Zanuy (EUPMt, Barcelona, Spain)
- Maria Giagkou (Institute for Language and Speech Processing, Greece)
- Milan Gnjatovič (University of Novi Sad, Serbia)
- Maurice Grenberg (New Bulgarian University, Sofia, Bulgaria)
- Jonathan Harrington (Ludwig-Maximilians University of Munich, Germany)
- Jennifer Hofmann (University of Zurich, Switzerland)
- Phil Hole (Ludwig-Maximilians University of Munich, Germany)
- Evgeniya Hristova (New Bulgarian University, Sofia, Bulgaria)
- Lazslo Hunyadi (University of Debrwcen, Hungary)
- Randy Klaassen (University of Twente, The Netherlands)
- Maria Koutsombogera (Institute for Language and Speech Processing, Greece)

- Barbara Lewandowska-Tomaszczyk (University of Lodz, Poland)
- Karmele Lopez-De-Ipina (Basque Country University, Spain)
- Saturnino Luz (University of Edinburgh, UK)
- Mauro Maldonato (Università della Basilicata, Italy)
- Rytis Maskeliunas (Kaunas University of Technology, Lithuania)
- Olimpia Matarazzo (Seconda Università di Napoli, Italy)
- Jiří Mekyska (Brno University of Technology, Czech Republic)
- Francesco Carlo Morabito, (Università "Mediterranea" di Reggio Calabria, Italy)
- Hossein Mousavi (Istituto Italiano di Tecnologia, Genova, Italy)
- Vittorio Murino (Istituto Italiano di Tecnologia, Genova, Italy)
- Costanza Navarretta (Centre for Language Technology, Njalsgade, Denmark)
- Rieks Op Den Akker (University of Twente, The Netherlands)
- Harris Papageorgiou (Institute for Language and Speech Processing, Greece)
- Eros Pasero (Politecnico di Torino, Italy)
- Jiří Přibil (Academy of Sciences, Czech Republic)
- Anna Přibilová (Slovak University of Technology, Slovakia)
- Zófia Ruttkay (Moholy-Nagy University of Art and Design Budapest, Hungary)
- Matej Rojc (University of Maribor, Slovenia)
- Michele Scarpiniti (Università di Roma "La Sapienza", Italy)
- Filomena Scibelli (Seconda Università di Napoli, Italy)
- Björn W. Schuller (Imperial College, UK and University of Passau, Germany)
- Zdenek Smékal (Brno University of Technology, Czech Republic)
- Jordi Solé-Casals (University of Vic, Spain)
- Stefano Squartini (Università Politecnica delle Marche, Italy)
- Igor Stankovic (University of Gothenburg, Sweden)
- Jing Su (Trinity College Dublin, Ireland)
- Jianhua Tao (Chinese Academy of Sciences, P.R. China)
- Alda Troncone (Seconda Università di Napoli and IIASS, Italy)
- Alessandro Vinciarelli (University of Glasgow, UK)
- Carl Vogel (Trinity College of Dublin, Ireland)
- Jerneja Žganec Gros (Alpineon, Development and Research, Slovenia)
- Paul A. Wilson (University of Lodz, Poland)
- B. Yegnanarayana (International Institute of Information Technology, Gachibowli, India)

Sponsoring Organizations

- Seconda Università di Napoli, Dipartimento di Psicologia
- International Institute for Advanced Scientific Studies "E.R. Caianiello" (IIASS, www.iiassvietri.it/), Italy
- Società Italiana Reti Neuroniche (SIREN, www.associazionesiren.org/)

Contents

About the Editors

Anna Esposito received her "Laurea Degree" *summa cum laude* in Information Technology and Computer Science from the Università di Salerno in 1989 with a thesis on: *The Behavior and Learning of a Deterministic Neural Net* (published on **Complex System**, vol 6(6), 507–517, **1992**). She received her PhD Degree in Applied Mathematics and Computer Science from the Università di Napoli, "Federico II" in 1995. Her PhD thesis was on: *Vowel Height and Consonantal Voicing Effects: Data from Italian* (published on **Phonetica**, vol 59(4), 197–231, **2002**) at Massachusetts Institute of Technology (MIT), Research Laboratory of Electronics (RLE), under the supervision of professor Kenneth N. Stevens.

She has been a Post Doc at the International Institute for Advanced Scientific Studies (IIASS), and Assistant Professor at the Department of Physics at the Università di Salerno (Italy), where she taught courses on Cybernetics, Neural Networks, and Speech Processing (1996–2000). She had a position as Research Professor (2000–2002) at the Department of Computer Science and Engineering at Wright State University (WSU), Dayton, OH, USA. She is currently associated with WSU as Research Affiliate.

Anna is currently working as an Associate Professor in Computer Science at the Department of Psychology, Seconda Università di Napoli (SUN). Her teaching responsibilities include Cognitive and Algorithmic Issues of Multimodal Communication, Human Machine Interaction, Cognitive Economy, and Decision Making. She authored 160+ peer reviewed publications in international journals, books and conference proceedings. She edited/co-edited 21 books and conference proceedings with Italian, EU and overseas colleagues.

Anna has been the Italian Management Committee Member of:

- COST 277: Nonlinear Speech Processing, http://www.cost.esf.org/domains_actions/ict/Actions/277 (2001–2005)
- COST MUMIA: **MUltilingual and Multifaceted Interactive information Access**, www.cost.esf.org/domains_actions/ict/Actions/IC1002 (2010–2014)
- COST TIMELY: Time in Mental Activity, www.timely-cost.eu (2010–2014)

She has been the proposer and chair of COST 2102: Cross Modal Analysis of Verbal and Nonverbal Communication, http://www.cost.esf.org/domains_actions/ict/Actions/2102 (2006–2010).

Since 2006, she is a Member of the **European Network for the Advancement of Artificial Cognitive Systems, Interaction and Robotics** (www.eucognition.org);

She is currently the Italian Management Committee Member of ISCH COST Action IS1406: Enhancing children's oral language skills across Europe and beyond (http://www.cost.eu/COST_Actions/isch/Actions/IS1406).

Anna's research activities are on the following three principal lines of investigations:

- **1998 to date**: Cross-modal analysis of speech, gesture, facial and vocal expressions of emotions. Timing perception in language tasks.
- **1995 to date**: Emotional and social believable Human-Computer Interaction (HCI).
- **1989 to date**: Neural Networks: learning algorithm, models and applications

Lakhmi C. Jain serves as a Visiting Professor in Bournemouth University, United Kingdom and Adjunct Professor in the Faculty of Education, Science, Technology and Mathematics in the University of Canberra, Australia. He is a Fellow of the Institution of Engineers Australia.

Dr. Jain founded the KES International for providing a professional community the opportunities for publications, knowledge exchange, cooperation and teaming. Involving around 5000 researchers drawn from universities and companies world-wide, KES facilitates international cooperation and generate synergy in teaching and research. KES regularly provides networking opportunities for professional community through one of the largest conferences of its kind in the area of KES.

www.kesinternational.org

His interests focus on the artificial intelligence paradigms and their applications in complex systems, security, e-education, e-healthcare, unmanned air vehicles and intelligent systems.

Chapter 1
More than the Modeling of Emotions: A Foreword

Leopoldina Fortunati

Abstract In this work I am providing comments on some important factors that must be accounted for when robotics comes near to the human body, as in the case of wearable robots. In addition to well-known socio-demographic factors such as sex/gender, age, ethnicity, skin color and emotional and culturally-determined sense of personal space, characteristics of health professionals and caregivers, here I stress the necessity to take into account also fashion and the questions raised by it.

Keywords Robotic assistive technologies · Mediated emotion · Fashion · Sociology of robotics

This volume represents a significant advancement of the knowledge required to build reliable autonomous systems in the various environments sustained by social robotics' products [1–6]. Its focus, that is on the modeling of emotions, stands on the ridge where the social sciences and hard sciences meet and integrate [7]. Emotions in fact are at the core of human relationships and processes such as socialization and education [8]. The study of emotion in itself is a challenge for anthropologists, sociologists, cognitive and social psychologists and neuroscientists, but when the emotions encounter technology and are filtered and mediated by some media, this challenge becomes even more problematic.

It is important to remember that a mediated emotion can be defined as "an emotion felt, narrated or showed, which is produced or consumed, for example in a telephone or mobile phone conversation, in a film or a TV programme or in a website, in other words mediated by a computational electronic device. Electronic emotions are emotions lived, re-lived or discovered through machines. Through ICT, emotions are on one hand amplified, shaped, stereotyped, re-invented and on the other sacrificed, because they must submit themselves to the technological limits and languages of a machine" [9]. What happens to emotions in the somersault towards technology is not only a technical problem, but it is a multidimensional problem. As Adam Smith pointed out in his first book *The Theory of Moral Sentiments* (1759) [10], emotions

L. Fortunati (✉)
Dipartimento di Studi Umanistici e del Patrimonio Culturale,
Università di Udine, Pordenone, Italy
e-mail: leopoldina.fortunati@uniud.it

© Springer International Publishing Switzerland 2016
A. Esposito and L.C. Jain (eds.), *Toward Robotic Socially Believable
Behaving Systems - Volume I*, Intelligent Systems Reference Library 105,
DOI 10.1007/978-3-319-31056-5_1

are the glue that keeps together the fabric of society. Advancing the understanding of emotion means to address *"the reasons of the heart"*, of which *"reason knows nothing"*, according to Blaise Pascal [11].

The focus of this volume is on the one hand on the social features of emotion in speech, handwriting, facial, vocal and gestural expressions. On the other, it is on people's ability to decode and encode emotional social cues in social interactions, which are a blend of emotion and reason. The findings collected here allow to elaborate multidimensional models of multimodal interactional features. These models will enable the design and development of social robots to be used, for example, in health care services, training/education systems, human–computer interfaces, entertainment, computer games, and communication and information spheres, last but not the least in the human body.

When robot technologies approach the human body, this becomes subject to a robotization process. We must be aware, in this case, of the social stratification and cultural layers that surround the human body. So far studies on robotics stress the importance to considerate a series of socio-demographic factors: sex/gender, age, ethnicity, skin color and emotional and culturally-determined sense of personal space, as well as characteristics of health professionals and caregivers.

Women and men often have similar needs, but not the same [12]. As an example, important for the development of social robotics, consider elders needs. Studies indicate that the interaction of sex and gender affects elderly's health [12]. Thus, understanding how sex and gender interact to affect the life conditions of old men and women is crucial to assist engineers in developing à la carte technologies that fit better their specific needs. Dementia strikes equally women and men, but as women live longer in most developed countries, these suffer more dementia [13, 14]. Arthritis and rheumatoid arthritis are more common in women than in age-matched men [15, 16]. Dexterity [17] and hearing impairment [18] impact men more than age-matched women. These differences depend on sex-specific biology, but also on other factors such as the different exposition to the occupational noise experienced more by men than by women [19].

For designing successful robotic assistive technologies it is important to be aware of sex and gender factors. Only taking into account these factors it is possible to design assistive technologies with a good marketability, usefulness, and acceptability. This will become even more important as the population continues to age. Data from Europe show that women represent a growing large proportion of older elderly. An important example of the influence of sex-gender factors is the partnering patterns, such as marriage age and age differences in partnerships [12]. In European Union countries and in the U.S., women tend to marry slightly older men [20, 21]. In Sweden and in Norway, age gaps are on average larger among homosexual than heterosexual couples [22–24]. Marriage age gaps that are a gendered phenomenon, when combined with women's greater longevity, means that (1) women are more likely to live alone than men; (2) women are more likely than men to be widowed, and (3) the death of a spouse is a major predictor of loneliness [25].

The combination of sex-gender factor may imply women have greater needs for assistive technologies that provide social connectivity. Overall, designers should be aware that gender differences in marriage age, partnering patterns, experience in household management, and receptivity to technology, are important to consider for effective design. At the same time they must acknowledge that the majority of the elderly are women and that women and men often have distinctive needs for physical mobility and cognitive dexterity. When addressing the market for robotic assistive technologies, researchers cannot but take into account women's and men's specific needs as elderly and as elder caregivers.

In developing emotional intelligence in robots, taking into account gendered differences in expression is crucial [26]. Moreover, other factors must be taken into account for robotic assistive technologies: ethnicity, skin color and emotional and culturally-determined sense of personal space. Regarding ethnicity, for example, current facial recognition algorithms—used to identify individuals, detect emotional cues, etc.—are often more effective in respect to subjects of one skin color than another [12]. In an international competition facial recognition algorithms developed by researchers in East Asian countries showed to be more accurate for Asian faces than Caucasian faces [27]. At the same time, algorithms developed in Western countries were more accurate for Caucasian faces than Asian faces. Including the factor of ethnicity in the design of robotic facial recognition systems for global markets might be very important. Finally, in the design if assistive robots also gendered and culturally-determined sense of personal space [28] cannot but be considered if engineers aim to build tools capable to interact with users in a socially-acceptable way.

Research and design of robot assistive technologies also gain by including the characteristics of health professionals and caregivers. In Europe, women are about twice as many men to provide informal care for ill or elderly adults [29]. In the U.S., about 70 % of informal care is provided by women [30, 31]. Over the years, the care givers develop hands-on knowledge that technology designers should access through participatory design [32]. This knowledge is very precious but designers need to consult also the persons with different relationships to elderly (sons, daughters, spouses, cousins, etc.) [33]. In Finland, researchers have studied both elderly's and caregivers' responses to over 60 assistive technologies in four pilot "smart homes" [34].

According to Schiebinger et al. [12], some assistive robots may work collaboratively with a smart home environment to address psychosocial conditions, such as isolation and depression. Robots may interact directly with users for a variety of purposes such as monitor their mental status, provide cognitive stimulation, offer companionship, and assist them with navigating complex environments [35]. To increase the acceptance of the user, some criteria for assistive technology used in physical interaction have been defined [36]. These criteria are embodiment, personality, empathy, engagement, adaption and transfer.

Women and men differ in their needs for and experience with technology. Women may have less technical experience and more positive attitudes toward technology [37]. They may also be more apprehensive about using assistive robots in domestic environments [38], but also more open than men to welcome them at home. Thus, it is important to include both women and men in the technology design.

Analyzing sex and gender as well as including both women and men users in technology development are positive actions that can lead to better designs and improve marketability of products. Researchers are developing new robot assistive technologies to support independent living for the elderly and to lighten the burdens of caregivers. Involving users and caregivers in the design process enhances outcomes. Thus, through participatory research and design with both the elderly and their caregivers, designers are in the position to gain key insights for developing assistive products that are useful to a broad user base.

However, in addition to these socio-demographic variables, I would like to stress here another factor that immediately comes into play when the human body is approached: fashion. Not by chance in 2001 I organized at the Triennale in Milano with a network of universities an important, interdisciplinary workshop on the relationship between the human body and technology attended by scholars coming from a variety of disciplines such as robotics, industrial design, medicine, sociology, communication, history of art, anthropology, psychology, fashion studies [39]. It was a first attempt to put many different disciplines around a table with the purpose to think and discuss how to integrate all these different glimpses into the same object of investigation and research: the human body in its encounter with technology.

One of the outcomes of this workshop was that fashion poses at least three questions.

First, when a technology comes around the human body it MUST come to terms with fashion. Fashion covers and manages the widest area of the human body. At the same time, it mediates fundamental things of everyday life, such the presentation of the self, the sense of beauty, etc. We learnt about this aspect with the mobile phones. All the other digital technologies such as television were a question of decor, how they could fit inside the decor of our apartment without disturbing it and integrating harmoniously inside. For example, one of the reasons of the delay on the part of Italians in the adoption of personal computers was that these devices at the beginning were so ugly that women could not decide where to place them in the house. Italian women did not perceive as possible to put the computer in the dining room. At the end, the children's bedroom was chosen in many homes to guest the computer, because there was a table (to study) and, at the same time, children's room was a space far from the social routes of the domestic environment. Compared to these technologies, the mobile phone developed a very different story. Very soon, this device from being considered a part of the equipment of a house, became a personal device. From that moment, the mobile phone stayed usually on the human body and from that moment, it became a question of fashion [37]. As consequence, it has had to be harmonized with the outfits of people. It could not remain a classical, ugly black box. It became cute and even more: fashionable. Two different worlds fought to have the predominance on the design of this device: the fashion system and the design field [40]. We have reasons to believe that also robots will need to become fashionable to stay on the human body. On the contrary, in the case they will be some sort of more traditional robots they will become immediately a question of décor. Where to put them inside the house will be the puzzle problem.

The second issue raised by fashion is the inevitable contamination that is produced by its encounter with technologies. In particular, for now on robotics is experimenting on materials and suits, and fashion is experimenting on robotics. The current experimentations need to converge, to fertilize each other. This process can be described as the "robotization" of fashion and the "fashionalizing" of robots. For the fashion system and for social robotics it is a very effervescent moment. Think for example to the integrated clothes systems for space crew [41], increasingly relevant field of research following the increase in the lengthening of space missions' duration. In addition, think for example also to so far unresolved communication issues connected to the health machines [42] and to the fear that their design aroused on patients. The long experimentation on "soft machines" [43] must be read as the attempt to overcome the sense of extraneousness felt by people towards the classic black box. In recent years, new textiles have begun to incorporate various kinds of micro- and nano-technologies and conductive materials able to react to changes in physical and environmental conditions. In this field of research, the "smart textiles" are engineered and support electronic circuits, micro-controllers, sensors, etc. They are not only functional; they become at the same time fashionable. Given the current prestige of technology in society, the dresses that highlight artificiality and look spectacular are attractive and fascinating. The reason is that they convey the sense of the inorganic and the imagery of the machine: the experimentation in this field goes in the direction to design clothes that can be seen as a kind of machines [44].

The Robot Companions for Citizens Manifesto launched by the Sant' Anna School contains many interesting proposals such the *"Robot suit that is a wearable robot that provides support to people when moving and doing everyday life activities"*. This is a genial idea and this will be a great part of the future innovation. We must pay attention to the already established field of "fashion tech", which has been around yet for many years. It may be sufficient to think about the research done and which represents the convergence of fashion design with engineering and participatory design experiences. As Danese [45] points out, this convergence can be exemplified in the projects carried out at the V2 Institute for the Unstable Media in Rotterdam, which recently hosted a lecture on 'Robotic Fashion and Intimated Interfaces', or in the 2012 'Technosensual' exhibition at the Wiener Museumquartier that was focused on technologically enhanced garments. Many of these experiments in the fashion tech try to shape intelligent systems around the human body and share many elements in common with the area of robotics related to the body, where the focus is often to enhance the wearability of the body-related devices. Here we cannot but mention the great, visionary work done by Hussein Challayan, the fashion designer of animatronic fashion.

The third issue posed by fashion is to consider if the body of robots when they enter in a home need to be dressed somehow. There are collections of possible clothes for robots [47]. At the same time there is the attempt to design 'fashionable' robots as in the case of robots designed by Simeon Gergiev for Highsnobiety (Givenchy Robotics) [48].

Although the chapters included in this volume represent surely a valuable progress in the direction of the modeling of emotions—necessary advancement for social robotics studies, it is necessary to acknowledge that we are far from having a multimodal system capable of working with spontaneous emotional material. The automatic synthesis and recognition of human emotions remains an example of computational task that cannot yet be solved by current pattern recognition and machine learning algorithms. To perform this task at a higher level it is necessary a stronger level of integration between hard and soft sciences. It is probably only by putting together these different areas of study that one can approach the complex nature of emotion. We are still far from a theory of emotion that encompasses this complexity, but at least there are several theories and approaches—like those presented in this volume—as well as many research and reflections on the emotional relationship between human beings and machines that can inspire the study of emotion [48–51].

References

1. Sugyiama S, Vincent J (eds) (2013) Special issue Social robots and emotion: transcending the boundary between humans and ICTs. intervalla: platform for intellectual exchange
2. Esposito A, Fortunati L, Lugano G (2014) Modeling emotion, behaviour and context in socially believable robots and ICT interfaces. Cogn Comput 6(4):623–627
3. Fortunati L, Esposito A, Ferrin G, Viel M (2014) Approaching social robots through playfulness and doing-it-yourself: children in action. Cogn Comput 6(4):789–801
4. Fortunati L, Esposito A, Lugano G (2015) Beyond Industrial robotics: social robots entering public and domestic spheres. Inf Soc: Int J 31(3):229–236
5. Fortunati L, Esposito A, Sarrica M, Ferrin G (2015) Children's knowledge and imaginary about robots. Int J Soc Robot 7(5):685–695
6. Vincent J, Taipale S, Sapio B, Lugano G, Fortunati L (eds) (2015) Social robots from a human perspective. Springer, Berlin
7. Fortunati L, Esposito A, Vincent J (2009) Introduction. In: Esposito A, Vich R (eds) Cross-modal analysis of speech, gestures, gaze and facial expressions. Springer, Berlin, pp 1–4
8. Vincent J, Fortunati L (eds) (2009) Electronic emotion. The mediation of emotion via information and communication technologies. Oxford, Peter Lang
9. Fortunati L, Vincent J (2009) Introduction. In: Vincent J, Fortunati L (eds) (2009) Electronic emotion. The mediation of emotion via information and communication technologies. Oxford, Peter Lang, pp 13–14
10. Smith A (1759) The theory of moral sentiments. http://www.marxists.org/reference/archive/smith-adam/works/moral/part07/part7d.htm Accessed 1 Aug 2008
11. O'Connell MR (1997) Blaise pascal: reasons of the heart. Eerdmans, Grand Rapid
12. Schiebinger L, Klinge I, Sánchez de Madariaga I, Schraudner M, Stefanick M (eds) (2011–2013). Gendered innovations in science, health & medicine, engineering and environment. http://ec.europa.eu/research/gendered-innovations/
13. Plassman B, Langa K, Fisher G, Herringa S, Weir D, Ofstedal M, Burke J, Hurd M, Potter G, Wodgers W, Steffens D, Willis R, Wallace R (2007) Prevalence of dementia in the united states: the aging, demographics, and memory study. Neuroepidemiology 29(1–2):125–132
14. Nowrangi M, Rao V, Lyketsos C (2011) Epidemiology, assessment, and treatment of dementia. J Psychiatr Clin N Am 34(2):275–294
15. Alamanos Y, Drosos A (2005) Epidemiology of adult rheumatoid arthritis. Autoimmun Rev 4(3):130–136
16. Linos A, Worthington J, O'Fallon W, Kurland L (1980) The epidemiology of rheumatoid arthritis in rochester, minnesota: a study of incidence, prevalence, and mortality. Am J Epidemiol 111(1):87–98

17. Desrosiers J, Hébert R, Bravo G, Dutil É (1995) Upper extremity performance test for the elderly (tempa): normative data and correlates with sensorimotor parameters. Arch Phys Med Rehabil 76(12):1125–1129

18. Cruickshanks K, Zhan W, Zhong W (2010) Epidemiology of age-related hearing impairment. In: Gordon-Salant T, Frisina R, Popper A, Fay R (ed) System the aging auditory, Springer Science and Business Media, New York, pp 259–274

19. Engdahl B, Krog N, Kvestad E, Hoffman H, Tambs K (2012) Occupation and the risk of bothersome tinnitus: results from a prospective cohort study (nord-trøndelag health study, hunt). Br Med J 2(1):e000512

20. Lakdawalla D, Schoeni R (2003) Is nursing home demand affected by the decline in age difference between spouses? Demograph Res 8(10):279–304

21. Van Poppel F, Liefbroer A, Vermunt J, Smeenk W (2001) Love, necessity, and opportunity: changing patterns of marital age hegemony in the netherlands, 1850–1993. Popul Stud: J Demogr 55(1):1–13

22. Andersson G, Noack T, Seierstad A, Weedok-Fekjær H (2006) The demographics of same-sex marriages in norway and sweden. Demography 43(1):79–98

23. Kristiansen J (2005) Age differences at marriage: the times, they are a changing? Statistics Norway, Oslo

24. Noack T, Seierstad A, Weedon-Fekjær H (2005) A demographic analysis of registered partnerships (legal same-sex unions): the case of norway. Eur J Popul 21(1):89–109

25. Dragset J, Kirkevold M, Espehaug B (2011) Loneliness and social support among nursing home residents without cognitive impairment: a questionnaire survey. Int J Nurs Stud 48(5):611–619

26. Phillips P, Jiang F, Narvekar A, Ayyad J, O'Toole A (2011) An other-race effect for face recognition algorithms. Assoc Comput Mach (ACM) Trans Appl Percept (TAP) 8(2):14–25

27. Brody L, Hall J (2008) Gender and emotion in context. In: Lewis M, Haviland-Jones J, Barrett L (eds) Handbook of emotions, 3rd edn. Guilford Press, New York, pp 395–408

28. Fries S (2005) Cultural, multicultural, cross-cultural, intercultural: a moderator's proposal. International Association of Teachers of English as a Foreign Language (IATEFL) Publications, Paris

29. Daly M, Rake K (2003) Gender and the welfare state: care, work, and welfare in Europe and the USA. Polity Press, Cambridge

30. Lahaie C, Earle A, Heymann J (2012) An uneven burden: social disparities in adult caregiving responsibilities, working conditions, and caregiver outcomes. Res Aging 35(3): 243–274

31. Kramer B, Kipnis S (1995) Eldercare and work-role conflict: toward an understanding of gender differences in caregiver burden. Gerontologist 35(3):340–348

32. Landau R, Auslander G, Werner S, Shoval N, Heinik D (2010) Families' and professional caregivers' views of using advanced technology to track people with dementia. Qual Health Res 20(3):409–419

33. Andersson S, Meiland F (2007) Cogknow field test #1 report. European Commission, Brussels

34. Melkas H (2013) Innovative assistive technology in finnish public elderly-care services: a focus on productivity work. J Prev Assess Rehabil 46(1): 77–91

35. Pollack M (2005) Intelligent technology for an aging population: the use of artificial intelligence (ai) to assist elders with cognitive impairment. AI Mag 26(2):9–24

36. Tapus A, Mataric' MJ, Scassellati B (2007) The grand challenges in socially assistive robotics. IEEE Robot Autom Mag 14(1):35–42

37. Fortunati L, Manganelli AM (1998) La comunicazione tecnologica: Comportamenti, opinioni ed emozioni degli Europei. In: Fortunati L (ed) Telecomunicando in Europa. Milano: Angeli, pp. 125–194

38. Cortellessa G, Scopelliti M, Tiberio L, Koch Svedberg G, Loutfi A, Pecora F (2008) A cross-cultural evaluation of domestic assistive robots. Proceedings of AAAI fall symposium on AI in eldercare: new solutions to old problems,

39. Fortunati L, Katz J, Riccini R (eds) (2003) Mediating the human body: technology. Communication and Fashion. Erlbaum, Mahwah

40. Fortunati L (2013) The mobile phone between fashion and design. Mob Media Commun 1(1):102–109
41. Dominoni A (2003) Conditions of microgravity and the body's "second skin". In: Fortunati L, Katz JE, Riccini R (eds) Mediating the human body: technology, communication and fashion, Erlbaum, Mahwah, pp 201–208
42. Chiapponi M (2003) Health care technologies. the contribution of industrial design. In Fortunati L, Katz JE, Riccini R (eds) Mediating the human body: technology, communication and fashion, Erlbaum: Mahwah, pp 187–194
43. Danese E (2003) Inside the surface. technology in modern textiles. In: Fortunati L, Katz JE, Riccini R (eds) Mediating the human body: technology, communication and fashion. Erlbaum, Mahwah, pp 147–154
44. Danese E. (2014) Fashion and technology. Paper presentation at the Doctoral Programme in Multimedia Communication. Pordenone, 27 marzo 2014
45. Danese E (2015) Fashion tech and robotics. In: Vincent J, Taipale S, Sapio B, Fortunati L, Lugano G (eds) Social robots from a human perspective. Springer, London, pp 129–138
46. http://robotswearingclothes.tumblr.com; http://www.cc.rim.or.jp/~comura/doll_web/life_size/lstd/lstd007.html
47. http://hypebeast.com/2014/7/givenchy-robotics-by-simeon-georgiev
48. Fortunati L (1995) Gli italiani al telefono. Angeli Editore, Milano
49. Fortunati L, Manganelli AM (2008) The social representations of telecommunications. Pers Ubiquitous Comput 12(6):421–431
50. Vincent J (2005) Are people affected by their attachment to their mobile phone? In: Nyiri K (ed) Communications in the 21st century. The global and the local in mobile communication: places, images, people, connection, Wien, Passagen Verlag
51. Vincent J (2006) Emotional attachment to mobile phones. Knowl Technol Policy 19:39–44

Chapter 2
Modeling Emotions in Robotic Socially Believable Behaving Systems

Anna Esposito and Lakhmi C. Jain

Abstract This book aims to investigate the features that are at the core of human interactions to model the involved emotional processes, in order to design and develop autonomous systems and algorithms able to detect early signs of changes, in moods and emotional states. The attention is focused on emotional social features and the human's ability to decode and encode emotional social cues while interacting. In order to do this, the book will propose a series of investigations that gather behavioral data from speech, handwriting, facial, vocal and gestural expressions. This is done through the definition of behavioral tasks that may serve to produce changes in the perception of emotional social cues. Specific scenarios are designed to assess users' emphatic and social competencies. The collected data are used to gain knowledge on how behavioral and interactional features are affected by individuals' moods and emotional states. This information can be exploited to devise multidimensional models of multimodal interactional features that will serve for measuring the degree of empathic relationships developed between individuals and allow the design and development of cost-effective emotion-aware technologies to be used in applicative contexts such as remote health care services and robotic assistance.

Keywords Socially believable robotic interfaces · Mood changes · Social and emotional interactional features · Speech · Gestures · Faces · Emotional expressions

A. Esposito (✉)
Dipartimento di Psicologia and IIASS, Seconda Università di Napoli, Caserta, Italy
e-mail: iiass.annaesp@tin.it

L.C. Jain
Bournemouth University, Bournemouth, UK
e-mail: jainlc2002@yahoo.co.uk

L.C. Jain
University of Canberra, Canberra, ACT 2601, Australia

© Springer International Publishing Switzerland 2016   9
A. Esposito and L.C. Jain (eds.), *Toward Robotic Socially Believable
Behaving Systems - Volume I*, Intelligent Systems Reference Library 105,
DOI 10.1007/978-3-319-31056-5_2

2.1   Introduction

The realization of a robotic agent capable of natural interactions with humans raises a number of issues and problems related to (a) the acquisition of sufficient competences to afford a full description and understanding of the user's perspective, and thus the creation of reliable and functional user models, (b) the agent's appearance and design, including the to be implemented interface, and (c) the reliability and credibility of a such artificial agent under the variety of application areas its tasks are devoted to [7, 12, 14]. The application areas may vary from training/education systems, to human-computer interfaces, entertainment, computer games, and several more. Each may require different agent's abilities, at a different level of complexity, ranging from natural speech communication to autonomous behaviors, detection and interpretation of user moods, personalities, and emotional behaviors, adaptation to the physical, organizational, and socio-cultural context, as well as, user needs and requirements.

Robotic socially believable behaving agents must be adaptive systems programmed to interact with humans by simulating typical spontaneous human interactions. To this aim it is needed a comprehensive model of the possible sets of "user" final conducts (the "user model"), as well as, in order for the agent to properly act, it is necessary to provide a proprioceptive description of the agent itself (the "model of itself", [5]. In human-machine interaction, a user model is what the system "knows " about the perspective and/or the expectations of the actual user, normally including both psychological data and a user profile that estimates the preferences of a specific class or a wide range of users. Psychological data tend to function effectively as a paradigm for a wider set of users, and are a reliable source for modeling the user's prospective [24]. For example, in the case of agents devoted to act a "butler" for guiding users in a foreign city, the user model needs to provide a comprehensive inventory of possible human requests and actions/reactions to expected and unexpected situations in terms of the reliability, confidence, facility of interaction/communication, as well as, trustworthiness, satisfaction, and credibility the users ascribe to the agent under different circumstances. Moreover, together a model of the possible user behavioral response, it is also important to anticipate such responses by means of a motivated design in order to minimize negative effects that can produce the user rejection of the agent.

Since emotions play a very important role in many aspects of our lives, including decision making, perception, learning, and behavior, and emotional skills are an important component of intelligence, especially for human-human interactions, the next generation of cognitive architectures must integrate and incorporate principles of emotionally colored interactions to define realistic models of human-machine interaction and suggest novel computational approaches for the implementation of real-time believable, autonomous, adaptive, and context-aware robotic agents [10, 13]. To this aim, fair efforts have been devoted to recognize and synthesize human emotional states exploiting different modalities, including: speech, facial expressions, gestures, and physiological signals as EEG, ECG, and others [1–3, 20–23, 26]. Some works investigated on the possibility to combine signals from multiple sources, with the general aim of improving the emotion classification accuracy and synthesis [6, 11]. However, a truly cognitive real-time multimodal system capable of working with spontaneous emotional material is still missing. The chapters proposed in this book aim to further progress in this direction, even though automatic synthesis and recognition of human emotions remain an example of computational task that cannot be perfectly solved by classical pattern recognition and machine learning algorithms.

2.2 Content of the Book

Sophisticated and functional computational instruments able to recognize, process, store, and synthesize relevant emotional signals, as well as interact with people, displaying reactions that show abilities of appropriately sensing and understanding (under conditions of limited time) environmental changes and producing suitable, autonomous, and adaptable responses created great promises in Information Communication Technology (ICT). However, Human Computer Interaction (HCI) paradigms that account for users' emotional states and emotional reactions are still far from being implemented. This book is signaling advances in these directions taking into account the multifunctional role of emotions in driving communication and interactional exchanges. To this aim, the book includes nine investigations on the role of emotions in shaping intentions, motivations, learning, and decision making. The first chapter by Vernon et al. [25] affords the problem of modeling (in socially believable behaving systems) actions, attentions, goals, and intentions, particularly intentions considered by the authors as the ability of the agent to read others minds and understand others' perspectives and beliefs. In this context, emotions will drive the efficient behavior of the agents, since as stated by the authors, the perception of social stimuli produces "*bodily states in the perceiving agen*t [and] *trigger* [in it] *affective states* [which in turn affect] *the agent's physical and cognitive performance*" [25, p. 3]. Corrigan et al. [8] faces similar problems from an implementation point of view considering aspects of human engagements with robotic and virtual agents, which are still actions guided by intentions and abilities to understand others' perspectives. The contribution of Belpaeme et al. [4] approaches similar problematics at a developmental level. The authors show that "*cognition emerges from the interaction between the brain, the body and the physical environment*" exploiting the iCub humanoid platform. The interesting paradigm emerging from their results is that "*artificial cognition, just as its natural counterpart, benefits from being grounded and embodied*" in a body, a brain, a physical, social context. The missing of one of these four constituents still allow to create artificial cognition, however, the complete system seems to be more efficient and effective. Meudt's et al. [19] contribution intend to show that the ability of an agent to recognize users' emotions will facilitate the users' adaptation process and improve the human interaction with the agent, making its function as companion or assistive technology more reliable. The contribution of Lewandowska-Tomaszczyk and Wilson [17] underline the physical and moral role of the disgust across different cultures, showing how important is the social setting and why *"an emotion-sensitive socially interacting robot would need to encode and decode* [such emotion] *in order to competently* and appropriately interact with the environmental culture [17, p. 1]. On a similar theme is the contribution of Maricchiolo et al. [18], which affords the analysis of nonverbal (gestural) and physiological (heart rate and skin conductance) reactions to disagreeable (disgusting?) messages. Surely very far from disgust is the contribution of Dupont et al. [9] which reports on the result of a four-year EU project investigation on the "laughter": the ILHAIRE Project (http://www.ilhaire.eu/project). The authors describe the collected data and the multi-determined role of the "laughter" in social interaction. Finally, the last two contributions of this book by Hunyadi et al. [16] and Gangamohan et al. [15] are dedicated to emotional speech and in particular to emotional prosodic features extracted from the HuComTech Corpus and all the emotional speech features able to recognize emotional vocal expressions.

2.3   Conclusions

The readers of this book will get a taste of the major research areas in modeling emotions and of the multifunctional role of emotions in generating actions, goals, attentions, and intentions, as well as on the different paradigms to model emotions by analyzing interactional exchanges. The research topics afforded by the book cover research fields related to psychology, sociology, philosophy, computer science, robotics, signal processing and human-computer interaction. The contributors to this volume are leading authorities in their respective fields. The book captures and presents interesting aspects of communicative exchanges and is fundamental in studies, such as robotics, where multidisciplinary facets need to be considered in order to succeed in the implementation of robotic companions, and assistive technologies. In particular the book covers aspects of emotional information processing during interactional exchanges which would lead to the implementation of socially believable, autonomous, adaptive, context-aware situated HCI systems.

References

1. Atassi H, Esposito A (2008) Speaker independent approach to the classification of emotional vocal expressions. In: Proceedings of IEEE conference on tools with artificial intelligence (ICTAI 2008), vol 1. Dayton, 3–5 Nov 2008, pp 487–494
2. Atassi H, Esposito A, Smekal Z (2011) Analysis of high-level features for vocal emotion recognition. In: Proceedings of 34th IEEE international conference on telecom and signal processing (TSP), Budapest, 18–20 Aug 2011, pp 361–366
3. Atassi H, Riviello MT, Smékal Z, Hussain A, Esposito A (2010) Emotional vocal expressions recognition using the cost 2102 italian database of emotional speech. In: Esposito A et al. (eds)

Development of multimodal interfaces: active listening and synchrony, LNCS 5967, Springer, Berlin, pp 255–267

4. Belpaeme T, Adams S, De Greeff J, Di Nuovo A, Morse A, Cangelosi A (2016) Social development of artificial cognition. This volume

5. Benyon D, Turner P, Turner S (2005) Designing interactive systems: people, activities, contexts, technologies. Pearson Education, Harlow

6. Castellano G, Kessous L, Caridakis G (2008) Emotion recognition through multiple modalities: face, body, gesture, speech. Affect and emotion in human-computer interaction. Springer, Berlin, pp 92–103

7. Cordasco G, Esposito M, Masucci F, Riviello MT, Esposito A, Chollet G, Schlögl S, Milhorat P, Pelosi G (2014) Assessing voice user interfaces: the vAssist system prototype. In: Proceedings of the 5th IEEE international conference on cognitive infocommunications, Vietri sul Mare, 5–7 Nov 2014, pp 91–96

8. Corrigan LJ, Peters C, Küster D, Castellano G (2016) Engagement perception and generation for social robots and virtual agents. This volume

9. Dupont S, Çakmak H, Curran W, Dutoit T, Hofmann J, McKeown G, Pietquin O, Platt T, Ruch W, Urbain J (2016) Laughter research: a review of the ILHAIRE project. This volume

10. Esposito A (2013) The situated multimodal facets of human communication. In Rojc M, Campbell N (Eds), Coverbal synchrony in human-machine interaction, chap. 7. CRC Press, Taylor & Francis Group, Boca Raton, pp 173–202

11. Esposito A, Esposito AM (2012) On the recognition of emotional vocal expressions: motivations for an holistic approach. Cogn Process 13(2):541–550

12. Esposito A, Fortunati L, Lugano G (2014) Modeling emotion, behaviour and context in socially believable robots and ICT interfaces. Cogn Comput 6(4):623–627

13. Esposito A, Esposito AM, Vogel C (2015) Needs and challenges in human computer interaction for processing social emotional information. Patter Recognit Lett 66:41–51

14. Fortunati L, Esposito A, Lugano G (2015) Beyond industrial robotics: social robots entering public and domestic spheres. Inf Soc: Int J 31(3):229–236

15. Gangamohan P, Kadiri SR, Yegnanarayana B (2016) Analysis of emotional speech: A review. This volume

16. Hunyadi L, István Szekrényes I, Kiss H (2016) Prosody enhances cognitive infocommunication: materials from the HuComTech corpus. This volume

17. Lewandowska-Tomaszczyk B, Wilson PA (2016) Physical and moral disgust with socially believable behaving systems in different cultures. This volume

18. Maricchiolo F, Gnisci A, Cerasuolo M, Ficca, Bonaiuto M (2016) Speaker's hand gestures can modulate receiver's negative reactions to a disagreeable verbal message. This volume

19. Meudt S, Schmidt-Wack M, Honold F, Schüssel F, Michael Weber M, Schwenker F, Palm G (2016) Going further in affective computing: how emotion recognition can improve adaptive user interaction. This volume

20. Milhorat P, Schlögl S, Chollet G, Boudyy J, Esposito A, Pelosi G (2014) Building the next generation of personal digital assistants. In: Proceedings of the 1st IEEE international conference on advanced technologies for signal and image processing - ATSIP'2014, Sousse, 17–19 March 2014, pp 458–463

21. Placidi G, Avola D, Petracca A, Sgallari F, Spezialetti M (2015) Basis for the implementation of an EEG-based single-trial binary brain computer interface through the disgust produced by remembering unpleasant odors. Neurocomputing 160:308–318

22. Ringeval F, Eyben F, Kroupi E, Yuce A, Thiran JP, Ebrahimi T, Lalanne D, Schuller B (2014) Prediction of asynchronous dimensional emotion ratings from audiovisual and physiological data. Pattern Recogn Lett Elsevier 66(C):22–30

23. Schuller B (2015) Deep learning our everyday emotions: a short overview. In: Bassis S et al (eds) Advances in neural networks: computational and theoretical issues, vol 37. SIST Series, Springer, Berlin, pp 339–346

24. van der Veer GC, Tauber MJ, Waern Y, van Muylwijk B (1985) On the interaction between system and user characteristics. Behav Inf Technol 4:284–308

25. Vernon D, Thill S, and Ziemke T (2016) The role of intention in cognitive robotics. This volume
26. Vinciarelli A, Esposito A, André E, Bonin F, Chetouani M, Cohn JF, Cristan M, Fuhrmann F, Gilmartin E, Hammal Z, Heylen D, Kaiser R, Koutsombogera M, Potamianos A, Renals S, Riccardi G, Salah AA (2015) Open challenges in modelling, analysis and synthesis of human behaviour in human-human and human-machine interactions. Cogn Comput 7(4):397–413

Chapter 3
The Role of Intention in Cognitive Robotics

D. Vernon, S. Thill and T. Ziemke

Abstract We argue that the development of robots that can interact effectively with people requires a special focus on building systems that can perceive and comprehend intentions in other agents. Such a capability is a prerequisite for all pro-social behaviour and in particular underpins the ability to engage in instrumental helping and mutual collaboration. We explore the prospective and intentional nature of action, highlighting the importance of joint action, shared goals, shared intentions, and joint attention in facilitating social interaction between two or more cognitive agents. We discuss the link between reading intentions and theory of mind, noting the role played by internal simulation, especially when inferring higher-level action-focussed intentions. Finally, we highlight that pro-social behaviour in humans is the result of a developmental process and we note the implications of this for the challenge of creating cognitive robots that can read intentions.

3.1 Introduction

There are many reasons why one would like a robot to exhibit a capacity for cognition. These include the ability to deal with uncertain or poorly specified situations and the ability to act prospectively, anticipating the need for actions and predicting the outcome of those actions [41, 42]. However, perhaps one of the most compelling motivations for research in cognitive robotics is the need for robots to interact naturally and safely with people. People are cognitive agents and consequently when

D. Vernon (✉) · S. Thill · T. Ziemke
School of Informatics, University of Skövde, Skövde, Sweden
e-mail: david.vernon@his.se

S. Thill
e-mail: serge.thill@his.se

T. Ziemke
Department of Computer and Information Science, Linköping University,
Linköping, Sweden
e-mail: tom.ziemke@his.se

© Springer International Publishing Switzerland 2016
A. Esposito and L.C. Jain (eds.), *Toward Robotic Socially Believable
Behaving Systems - Volume I*, Intelligent Systems Reference Library 105,
DOI 10.1007/978-3-319-31056-5_3

they interact socially with other agents, they exhibit the key attributes of cognition: prospection and adaptive real-time goal-directed behaviour. To interact effectively, a cognitive robot needs to be able to interact on the same basis. When that interaction stretches to helping or collaborating with people, a cognitive robot needs to be able to engage in perspective-taking, i.e. to form a theory of mind [25], to see the world from the other person's perspective.

In this chapter, we first explore the four attributes of cognition involved in inter-action with an inanimate world: attention, action, goals, and intentions, focusing in particular on the pivotal role of intention. We then expand the discussion to include interaction with other cognitive agents and discuss how these four attributes are extended in social interaction in the form of joint attention, joint action, shared goals, and shared intentions. In doing this, we explain that shared intention involves more that just the superposition of the intentions of two or more individual agents and highlight the essential role it plays in facilitating safe, engaging, and effective interaction, particularly where two or more agents are helping each other in fulfilling some task. By extension, we argue that cognitive robots as much as people need to have a capacity for reading and sharing intentions when interacting with people if they are to do so effectively [3].

3.2   The Prospective and Intentional Nature of Action

The movements of cognitive agents are organized: they are defined by goals and guided by prospection. These goal-directed prospectively-controlled move-ments are called actions [19, 42]. Typically, cognitive agents do not deliberatively pre-select the exact movements required to achieve a desired goal. Instead, they select prospectively-guided intention-directed goal-focussed action, with the spe-cific movements being adaptively controlled as the action is executed.

While action and goals are two of the essential characteristics of cognitive inter-action, there are two more: intention and attention. The first of these—intention—captures the prospective nature of action and goals. The distinction between inten-tions and goals is not always clearly made. An intention can be viewed as a plan of action an agent chooses and commits itself to in pursuit of a goal. An intention there-fore includes both a goal and the means of achieving it [8, 40]. Thus, an agent may have a goal for some state of affairs to exist and an intention to do something specific in pursuit of that state of affairs. Intentions integrate, in a prospective framework, actions and goals. Finally, there is perception, the essential sensory aspect of cogni-tion. However, in the context of cognitive action, perception is directed. It is focussed on goals and influenced by expectations. In other words, it is attentive. Arguably, one can describe attention in the context of interaction as an intention-guided perception [41].

All of the components of cognitive interaction—action, goals, intention, and attention—have an element of prospection. Our aim in this chapter is to explain what is necessary to transform this characterization to one that is representative of social interaction between two (or more) cognitive agents. This will involve the notions of *joint action*, *shared goals*, *shared intentions*, and *joint attention*. As we will see, this transformation, and these four notions, go beyond a simple superposition of individual action, goals, intention, and attention from which they derive. Much more is involved in social cognition and the interaction of two or more agents. To set the scene for this, we begin with a brief overview of social cognition and social interaction.

3.3   Social Cognition and Social Interaction

Social cognition—necessary for effective social interaction with other cognitive agents—embraces a wide range of topics. The abilities required for successful social interaction include reading faces, detecting eye gaze, recognizing emotional expressions, perceiving biological motion, paying joint attention, detecting goal-directed actions, discerning agency, imitation, deception, and empathy, among many others [15].

3.3.1   The Basis of Social Cognition

Social cognition depends on an agent's ability to interpret a variety of sensory data that conveys information about the activities and intentions of other agents. Newborns have an innate sensitivity to biological motion [37] and it has been shown that the ability to process biological motion is a hallmark of social cognition, providing a cognitive agent with a capacity for adaptive social behaviour and nonverbal communication, to the extent that individuals who exhibit a deficit in visual processing of biological motion are also compromised in social perception [33]. The clearest example of this is the ability to read body language, the subtle body movements, gestures, and actions that are an essential aspect of successful interaction between cognitive agents.

For an agent to interact socially with another cognitive agent, it must be (and stay) attuned to the cognitive state of that agent and be sensitive to changes. There is a strong link between the state of an agent's body and its cognitive and affective state, especially during social interaction [1, 22]. There are four aspects to this link:

1. When an agent perceives a social stimulus, this perception produces bodily states in the perceiving agent.
2. The perception of bodily states in other agents frequently evokes a tendency to mimic those states.
3. The agent's own body states trigger affective states in the agent.
4. The efficiency of an agent's physical and cognitive performance is strongly affected by the compatibility between its bodily states and its cognitive states.

Because of the link between bodily states and cognitive and affective states, the posture, movements, and actions of an agent convey a great deal about the agent's cognitive and affective disposition as well as influencing how another agent behaves towards it.

3.3.2 Helping and Collaboration

While social cognition is ultimately about mutual interaction, this interaction can be asymmetric or symmetric: one agent can assist another, or both agents can assist each other. In the following, we will refer to these behaviours as helping (sometimes adding the qualification *instrumental* helping) and collaboration, respectively. Significantly, the development of a capacity for collaborative interaction depends on the prior development of a capacity for instrumental helping and it takes several years for human infants to develop the requisite abilities [28].

During the first year of life the progressive acquisition of motor skills determines the development of the ability to understand the intentions of other agents, from anticipating the goal of simple movements to the understanding of more complex goals [14, 16]. At the same time, the ability to infer what another agent is focussing their attention on and the ability to interpret emotional expressions begins to improve substantially [10, 17].

At around 14–18 months of age children begin to exhibit instrumental helping behaviour, i.e. they display spontaneous, unrewarded helping behaviours when another person is unable to achieve his goal. Young children are naturally altruistic and have an innate propensity to help others instrumentally, even when no reward is offered [43]. This is a critical stage in the development of a capacity for collaborative behaviour, a process that progresses past three and four years of age.

Instrumental helping has two components: a cognitive one and an emotional one. The cognitive component is concerned with recognizing what the other agent's goal is: what they are trying to do. The motivational component is what drives the helping agent to act in the first place. This could be the desire to see the second agent achieve the goal or, alternatively, the desire to see the second agent exhibit pleasure at achieving the goal.

The ability to engage in instrumental helping develops with age: 14-month-old infants can help others in situations where the task is relatively simple, e.g. helping with out-of-reach objects, whereas 18-month-old infants engage in instrumental helping in situations where the cognitive task is more complicated [43]. As already mentioned, rewards are not necessary and the availability of rewards does not increase the incidence of helping. Indeed, rewards can sometimes undermine the motivation to help. Infants are willing to help several times and will even continue to help even if the cost of helping is increased.

The second form of helping—collaboration—is more complicated and focuses on mutual helping where two agents work together to achieve a common goal. It requires the two agents to share their intentions, to agree on the goal, share attention, and engage in joint action. Collaboration requires complex interaction over and above the ability to engage in instrumental helping. It involves the establishment of shared goals and shared intentions and it requires subtle adjustment of actions when the two agents are in physical contact such as when handing items to each other or carrying objects together. Michael Tomasello and Malinda Carpenter argue that *shared intentionality*, i.e. a collection of social-cognitive and social-motivational skills that allow two or more participants engaged in collaborative activity to share psychological states with one another, plays a crucial role in the development of human infants. In particular, it allows them to transform an ability to follow another agent's gaze into an ability to jointly pay attention to something, to transform social manipulation into cooperative communication, group activity into collaboration, and social learning into instructed learning [39].

3.3.3   The Central Role of Intention in Mutual Interaction

The situation becomes complicated when one progresses from instrumental helping to collaboration. In the latter case, we are dealing with *joint cooperative action*, or *joint action* for short, sometimes referred to as *shared cooperative activity*. Agents that engage in joint action share the same goal, intend to act together, and coordinate their actions to achieve their shared goal through joint attention. That sounds fairly straightforward but as we unwrap each of these issues—joint action, shared intentions, shared goals, and joint attention—interdependencies between them arise. For example, joint action requires a shared intention, a shared goal, and joint attention when executing the joint action; shared intention includes shared goals; and joint attention is effectively perception that is guided by shared intention and is goal-directed (see Fig. 3.1).

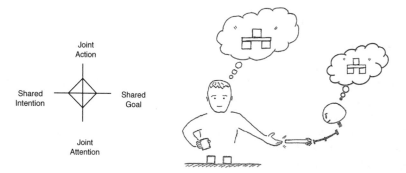

Fig. 3.1 Collaboration involves joint action, shared intention, a shared goal, and joint attention, each of which are mutually dependent. Here the human and the robot are engaged in a joint action and have a shared intention (and hence a shared goal and commitment to achieving it) and exhibit joint attention where their goal-directed perceptions are guided by their shared intention

For one agent to be able to help another agent, it must first infer or read the other agent's intentions. This in itself is a complex problem. It can be addressed in two phases: reading low-level intentions associated with movements (e.g. predicting what someone is reaching for) and reading high-level intentions associated with actions (e.g. predicting why someone is reaching for that object or what he or she want to do with it). Elisabeth Pacherie argues that three different levels of intentions can be distinguished: (1) distal intentions, where the goal-directed action to be performed is specified in cognitive terms with reference to the agent's environment; (2) proximal intentions, where the action is specified in terms of bodily action and the associated perceptual consequence; and (3) motor intentions, where the action is specified in terms of the motor commands and the impact on the agent's sensors [30].

To take part in collaborative activities requires an ability to read intentions and infer goals (as was the case in instrumental helping) but it also requires a unique motivation to share psychological states with other agents. By shared intentionality we here mean, following Tomasello et al., "collaborative actions in which participants have a shared goal (shared commitment) and coordinated action roles for pursuing that shared goal" [40]. What is significant is that the goals and intentions of each agent involved in the collaboration must include something of the goals and intentions of the other agent and something of its own goals and intentions. In other words, the intention is a joint intention and the associated actions are joint actions. This differentiates collaboration from instrumental helping and, as we have said, makes it more complicated. Furthermore, each agent understands both roles of the interaction and so can help the other agent if required. Critically, agents not only choose their own action plan, but also represent (or 'mirror') the other agent's action plan in its own motor system to enable coordination in the sense of who is doing what and when.

Assuming collaboration on a shared goal, let us now look more closely at the issues of joint action, shared intention, and joint attention to better understand the role of intention in the interaction between cognitive agents, be they human or robot.

3.3.4 Joint Action

There are at least six degrees of freedom in joint action [31]. These include the number of participants involved, the nature of the relationship between the participants (e.g. peer-to-peer or hierarchical), whether or not the roles are interchangeable, whether the interaction is physical or virtual, whether or not the participants' association is temporary or more long-lasting, and whether or not the interaction is regulated by organizational or cultural norms. In the following, we will assume physical joint action between two peers that temporarily collaborate on a shared goal.

According to Michael Bratman, joint action, or shared cooperative activity, has three essential characteristics [7]:

1. Mutual responsiveness;
2. Commitment to joint activity;
3. Commitment to mutual support.

Let's assume there are two agents engaged in a shared cooperative activity. Each agent must be mutually responsive to the intentions and actions of the other and each must know that the other is trying to be similarly responsive. Consequently, each agent behaves in a way that is guided partially by the behaviour of the other agent. This is different from instrumental helping where the helping agent is responsive to the intentions of the agent that needs help but not the other way round.

Each agent must also be committed to the activity in which they are engaged. This means that both agents have the same intention but they need not have the same reason for engaging in the activity. This is a subtle point: it means that the outcome of the collaboration is the same for both agents but the reason for adopting the goal of achieving that outcome need not be the same. If a cognitive robot and a disabled person collaborate to do the laundry, the outcome—the goal—may be a wardrobe full of clean clothes but the reason the person has the goal is to have a fresh shirt to wear in the morning whereas the reason the robot has the goal may just be to keep the house clean and uncluttered. If they collaborate to cook a stew, the goal may be nutritious meal, but the person's reason for the goal is to stay healthy whereas the reason the robot adopts the goal may simply be to use up some vegetables that would otherwise have to be thrown out.

Finally, each agent must be committed to supporting the efforts of the other to play their role in the joint activity. This characteristic complements the mutual responsiveness by requiring that each agent will in fact provide any help the other agent requires. It says that each agent treats this collaborative mutual support as a priority activity: even if there are other activities that are competing for the attention of each agent, they will still pay attention to the shared cooperative activity they are both engaged in.

Philip Cohen and Hector Levesque address similar issues in their theory of teamwork [11]. They do so in the context of designing artificial agents that can engage in joint action, setting out the conditions that need to be fulfilled for a group of agents to exhibit joint commitment and joint intention. This builds on their definitions of individual commitment and individual intention which are, roughly speaking, a persistent goal and an action plan in support of achieving that persistent goal.

It is important to note that Bratman's account of joint action has been subject to some criticism in that it appears to require sophisticated shared intentionality and an adult-level theory of mind. Yet, as we have seen, young children develop a capability for joint action. An alternative account that doesn't require sophisticated shared intentionality, but only requires shared goals and an understanding of goal-directed actions has been proposed; see [2, 9, 32]. That said, shared intentions are important for joint action and the intentions of each agent must interlock: each agent must intend that the shared activity be fulfilled in part by the other agent and that their individual activities—both planned actions and actual actions when being executed—mesh together in a mutually-supportive manner.

3.3.5 Shared Intentions

A shared intention—sometimes called *we-intention, collective intention,* or *joint intention*—is not simply a collection of individual intentions, even when those individual intentions are supplemented by beliefs or knowledge that both participating agents share [39]. There is more to it than this.

An agent with an individual intention represents the overall goal and the action plan by which it will achieve that goal and, furthermore, this plan is to be performed by the agent alone. That much is clear. However, agents with a shared intention (and engaged in a joint action) represent the overall shared goal between them but only their own partial sub-plans. Elisabeth Pacherie identifies three levels of shared intentions (shared distal intentions, shared proximal intentions, and coupled motor intentions) [31]; these are extensions of her characterization of individual intentions [30] (Sect. 3.3.3).

Shared intentionality appears to be unique to humans and its development seems to depend on a peculiarly-human motivation to share emotions, experience, and activities and a more general motivation to understand others as animate, goal-directed, and intentional [40]. An example of how artificial cognitive systems can exploit these ideas can be found in Peter Ford Dominey's and Felix Warneken's paper "The basis of shared intentions in human and robot cognition." Based on findings in computational neuroscience (e.g. the mirror neuron system) and developmental psychology, it describes how representations of shared intentions allow a robot to cooperate with a human [13].

Each individual agent with a shared intention does not need to know the other agent's partial plan. However, they do need to share the overall goal. When it comes to the realization of a shared intention and the execution of a joint action, the agent must also factor in the real-time coordination of their individual activities. In this case, each agent must also represent its own actions and their predicted consequences *and* the goals, intentions, actions and predicted consequences of the other agent [36]. Furthermore, each agent must represent the effect that their actions have on the other agent, it must have at least a partial representation of how component actions combine to achieve the overall goal, it must be able to predict the effects of their joint actions so that it can monitor progress towards the overall goal and adjust its actions to help the other agent if necessary. The additional requirements imposed by the execution of joint action correspond to Elisabeth Pacherie's shared proximal intentions [31].

It is apparent that, in carrying out a joint intention and executing a joint action, both agents must establish a shared perceptual framework. This is where joint attention (in the sense of perception guided by shared intention) comes in.

3.3.6 Joint Attention

Social interaction, in general, and collaborative behaviours, in particular, depend on the participating agents to establish *joint attention* [21]. Joint attention involves much more than two agents looking at the same thing. As Michael Tomasello and Malinda

Carpenter note, joint attention "is not just two people experiencing the same thing at the same time, but rather it is two people experiencing the same thing at the same time and *knowing together that they are doing this*" [39]. The essence of joint attention lies in the relationship between intentionality and attention. This provides the basis for a definition of joint attention as "(1) a coordinated and collaborative coupling between intentional agents where (2) the goal of each agent is to attend to the same aspect of the environment [21]". Joint attention, then, requires shared intentionality. Furthermore, the participating agents must be engaged in collaborative intentional action. During this collaboration, each agent must monitor, understand, and direct the attentional behaviour of the other agent, and significantly, both agents must be aware that this is going on.

Joint attention is an on-going mutual activity that is carried on throughout the collaborative process to monitor and direct the attention of the other agent. In a sense, joint attention is, itself, a joint activity.

At least four skills need to be recruited by a cognitive agent to achieve joint attention [21]. First, the agent must be able to detect and track the attentional behaviour of the other agent (we are assuming that there are just two agents involved in joint attention here but of course there could be more). Second, the agent must be able to influence the attentional behaviour of the other agent, possibly by using gestures such as pointing or by use of appropriate words. Third, the agent must be able to engage in social coordination to manage the interaction, using techniques such as taking turns or swapping roles, for example. Finally, the agent must be aware that the other agent has intentions (which, as we noted, could be different provided the goal is the same). That is, the agent must be capable of intentional understanding: it must be able to interpret and predict the behaviour of the other agent in terms of the actions required to reach the shared goal.

3.4   Reading Intentions and Theory of Mind

The ability to infer intentions is closely linked to what is known as *theory of mind* [23]: the capacity by which one agent is able to take a perspective on someone else's situation. Theory of mind is defined by Andrew Meltzoff as "the understanding of others as psychological beings having mental states such as beliefs, desires, emotions, and intentions" [23]. To have a theory of mind means to have the ability to infer what someone else is thinking and wants to do. The ability to imitate—the capacity to learn new behaviours by observing the actions of others [4, 24]—forms the basis for the development of a person's ability to form a theory of mind [25]. It is a key mechanism in cognitive development and it is innate in humans [26, 27].

The link between imitation and theory of mind is the ability of an agent to infer the intentions of another agent. When imitating adults, infants as young as 18 months of age can not only replicate the actions of the adult (and remember: actions are focussed on goals, not just bodily movements) when successfully performing a task but they can also persist in trying to achieve the goal of the action even when the adult is unsuccessful in performing the task. In other words, the infant can read the intention of the adult and infer the unseen goal implied by the unsuccessful attempts. Andrew Meltzoff and Jean Decety summarize the link between imitation and theory of mind (which they also refer to as *mentalizing*) as follows: "Evidently, young toddlers can understand our goals even if we fail to fulfil them. They choose to imitate *what we meant to do*, rather than what we mistakenly did do" [25], p. 496 (emphasis added). They also remark that "In ontogeny, infant imitation is the seed and the adult theory of mind is the fruit."

Young children normally differentiate between the behaviour of inanimate and animate objects, attributing mental states to the animate objects. In fact, such is the importance of biological motion to social cognition that if an inanimate object, even two-dimensional shapes such as triangles, exhibit movements that are animate or biological—self-propelled, non-linear paths with sudden changes in velocity— humans cannot resist attributing intentions, emotions, and even personality traits to that inanimate object [18]. In the same way, humans also infer different types of intention depending on whether they are interpreting movements (lower level intentions) or actions (higher level). Whereas movement intention refers to *what* physical state is intended by a certain action, e.g., inferring the end location of a specific observed movement—if the hand moves into the direction of a cup, it is likely that the agent intends to grasp that cup—a higher conceptual level intention refers to *why* that specific action is being executed and the motives underlying the action, e.g., the agent might be thirsty and want a drink. This mirrors the distinction we drew at the beginning between the concrete movements comprising an action and the higher-order conceptual goals of an action.

So, how do humans infer the intentions of others from their actions? Internal simulation is a possible mechanism [5].[1] The key idea is that the ability to infer the intentions of another agent from observations of their actions might actually be based on the same mechanism that predicts the consequences of the agent's own actions based on its own intentions. Cognitive systems make these predictions by internal simulation using forward models that take either overt or covert motor commands as input and produce as output the likely sensory consequences of carrying out those commands. When a cognitive system observes another agent's actions, the same mechanism can operate provided that the internal simulation mechanism is able to associate observed movements (and not just self-generated motor commands) and likely, i.e. intended, sensory consequences. This is what the ideo-motor principle suggests [20, 29, 38] and what the mirror-neuron system provides [34, 35]. By exploiting internal simulation, when an agent just sees another agent's action, not only are the actions activated in it but so too are the consequences of those actions, and hence the intention of the actions can be inferred. With a suitably-sophisticated joint representation and internal simulation mechanism, both low-level movement intentions and high-level action intentions can be accommodated.

---

[1]For an in-depth discussion of a computational approach to intention recognition, see "Towards computational models of intention detection and intention prediction" by Elisheva Bonchek-Dokow and Gal Kaminka [6].

Predicting and recognizing intentions in situations where there are groups of agents is particularly challenging because the cognitive system has to do more than track and predict the actions of individual agents, it also has to infer the joint intention of the entire group and this may not simply be "the sum of the intentions of the individual agent" [12]. It is also necessary to recognize the position of each agent in the social structure of the group. Again, this is a difficult challenge because an agent may play more than one role in a group.

3.5   Conclusions

Our goal in this chapter has been to highlight the pivotal role played by intentions in social interaction and, in particular, to argue that a capacity to infer the intentions of the cognitive agent with which one is interacting is essential if that interaction is to be effective. This is true both in the asymmetric case of instrumental helping where one agent assists another to achieve its goals without implicit instruction to do so and also in the symmetric case where both agents are collaborating. In this latter case, the ability to read intentions, by taking a perspective on the other agent's view of the interaction and forming a theory of mind for that agent, is doubly important because the other components of successful collaborative interaction—joint action and joint attention—also depends on the other agent's intentions. The upshot of this is that if we seek to construct cognitive robots that can interact effectively—asymmetrically or symmetrically: helping or collaborating—then it is imperative that these cognitive robots have a capacity for perspective taking and forming a theory of mind. However, it is not simply a question of implanting such a capacity in a cognitive robot. We know from psychology that this capacity is the result of an extended period of cognitive and social development where the emergence of the ability to engage in instrumental helping precedes that of collaborative interaction. Consequently, the challenge in cognitive robotics is to model the developmental process by which these capacities emerge over time as the robot engages with people.

Acknowledgments This work was supported by the European Commission, Project 611391 DREAM: Development of Robot-enhanced Therapy for Children with Autism Spectrum Disorders (www.dream2020.eu). Also supported by the Knowledge Foundation, Stockholm, under SIDUS grant AIR (Action and intention recognition in human interaction with autonomous systems). The authors would like to acknowledge the work of Harold Bekkering, Radboud University Nijmegen, in our characterization of aspects of neonatal development.

References

1. Barsalou LW, Niedenthal PM, Barbey A, Ruppert J (2003) Social embodiment. In: Ross B (ed) The psychology of learning and motivation, vol 43. Academic Press, San Diego, pp 43–92
2. Bekkering H, de Bruijn E, Cuijpers R, Newman-Norlund RD, van Schie H, Meulenbroek R (2009) Joint action: neurocognitive mechanisms supporting human interaction. Top Cogn Sci 1(2):340–352

3. Bicho E, Erlhagen W, Sousa E, Louro L, Hipolito N, Silva EC, Silva R, Ferreira F, Machado T, Hulstijn M, Maas Y, de Bruijn ERA, Cuijpers RH, Newman-Norlund RD, van Schie HT, Meulenbroek RGJ, Bekkering H (2012) The power of prediction: robots that read intentions. In: Proceedings of the IEEE/RSJ international conference on intelligent robots and systems, Vilamoura, Algarve, pp 5458–5459
4. Billard A (2002) Imitation. In: Arbib MA (ed) The handbook of brain theory and neural networks. MIT Press, Cambridge, pp 566–569
5. Blakemore S, Decety J (2001) From the perception of action to the understanding of intention. Nat Rev Neurosci 2(1):561–567
6. Bonchek-Dokow E, Kaminka GA (2014) Towards computational models of intention detection and intention prediction. Cogn Syst Res 28:44–79
7. Bratman ME (1992) Shared cooperative activity. Philos Rev 101(2):327–341
8. Bratman ME (1998) Intention and personal policies. In: Tomberlin JE (ed) Philosophical perspectives, vol 3. Blackwell, London
9. Butterfill S (2012) Joint action and development. Philos Q 62(246):23–47
10. Butterworth G, Jarrett N (1991) What minds have in common is space: spatial mechanisms serving joint visual attention in infancy. Br J Dev Psychol 9:55–72
11. Cohen PR, Levesque HJ (1991) Teamwork. Nous 25(4):487–512
12. Demiris Y (2007) Prediction of intent in robotics and multi-agent systems. Cogn Process 8:152–158
13. Dominey PF, Warneken F (2011) The basis of shared intentions in human and robot cognition. New Ideas Psychol 29:260–274
14. Falck-Ytter T, Gredebäck G, von Hofsten C (2006) Infants predict other people's action goals. Nat Neurosci 9(7):878–879
15. Frith U, Blakemore S (2006) Social cognition. In: Kenward M (ed) Foresight cognitive systems project, Foresight Directorate, Office of Science and Technology, 1 Victoria Street, London, SW1H 0ET. United Kingdom
16. Gredebäck G, Kochukhova O (2010) Goal anticipation during action observation is influenced by synonymous action capabilities, a puzzling developmental study. Exp Brain Res 202(2):493–497
17. Harris PL (2008) Children's understanding of emotion. In: Lewis M, Haviland-Jones JM, Barrett LF (eds) Handbook of emotions. The Guilford Press, New York, pp 320–331
18. Heider F, Simmel M (1944) An experimental study of apparent behaviour. Am J Psychol 57:243–249
19. von Hofsten C (2004) An action perspective on motor development. Trends Cogn Sci 8:266–272
20. Iacoboni M (2009) Imitation, empathy, and mirror neurons. Annu Rev Psychol 60:653–670
21. Kaplan F, Hafner V (2006) The challenges of joint attention. Interact Stud 7(2):135–169
22. Lindblom J (2015) Embodied social cognition, cognitive systems monographs (COSMOS), vol 26. Springer, Berlin
23. Meltzoff AN (1995) Understanding the intentions of others: re-enactment of intended acts by 18-month-old children. Dev Psychol 31:838–850
24. Meltzoff AN (2002) The elements of a developmental theory of imitation. In: Meltzoff AN, Prinz W (eds) The imitative mind: development, evolution, and brain bases. Cambridge University Press, Cambridge, pp 19–41
25. Meltzoff AN, Decety J (2003) What imitation tells us about social cognition: a rapprochement between developmental psychology and cognitive neuroscience. Philos Trans R Soc Lond: Ser B 358:491–500
26. Meltzoff AN, Moore MK (1977) Imitation of facial and manual gestures by human neonates. Sci 198:75–78
27. Meltzoff AN, Moore MK (1997) Explaining facial imitation: a theoretical model. Early Dev Parent 6:179–192
28. Meyer M, Bekkering H, Paulus M, Hunnius S (2010) Joint action coordination in 2-and-a-half- and 3-year-old children. Front Hum Neurosci 4(220):1–7

29. Ondobaka S, Bekkering H (2012) Hierarchy of idea-guided action and perception-guided movement. Front Cogn 3:1–5
30. Pacherie E (2008) The phenomenology of action: a conceptual framework. Cogn 107:179–217
31. Pacherie E (2012) The phenomenology of joint action: self-agency vs. joint-agency. In: Seemann A (ed) Joint attention: new developments. MIT Press, Cambridge, pp 343–389
32. Pacherie E (2013) Intentional joint agency: shared intention lite. Synth 190(10):1817–1839
33. Pavlova MA (2012) Biological motion processing as a hallmark of social cognition. Cereb Cortex 22:981–995
34. Rizzolatti G, Craighero L (2004) The mirror neuron system. Annu Rev Physiol 27:169–192
35. Rizzolatti G, Fadiga L (1998) Grasping objects and grasping action meanings: the dual role of monkey rostroventral premotor cortex (area F5). In: Bock GR, Goode JA (eds) Sensory guidance of movement, novartis foundation symposium, vol 218. Wiley, Chichester, pp 81–103
36. Sebanz N, Knoblich G (2009) Prediction in joint action: what, when, and where. Top Cogn Sci 1:353–367
37. Simion F, Regolin L, Bulf H (2008) A predisposition for biological motion in the newborn baby. Proc Natl Acad Sci (PNAS) 105(2):809–813
38. Stock A, Stock C (2004) A short history of ideo-motor action. Psychol Res 68(2–3):176–188
39. Tomasello M, Carpenter M (2007) Shared intentionality. Dev Sci 10(1):121–125
40. Tomasello M, Carpenter M, Call J, Behne T, Moll H (2005) Understanding and sharing intentions: the origins of cultural cognition. Behav Brain Sci 28(5):675–735
41. Vernon D (2014) Artificial cognitive systems – a Primer. MIT Press, Cambridge
42. Vernon D, von Hofsten C, Fadiga L (2010) A roadmap for cognitive development in humanoid robots, cognitive systems monographs (COSMOS), vol 11. Springer, Berlin
43. Warneken F, Tomasello M (2009) The roots of human altruism. Br J Psychol 100(3):455–471

Chapter 4
Engagement Perception and Generation for Social Robots and Virtual Agents

Lee J. Corrigan, Christopher Peters, Dennis Küster
and Ginevra Castellano

Abstract Technology is the future, woven into every aspect of our lives, but how are we to interact with all this technology and what happens when problems arise? Artificial agents, such as virtual characters and social robots could offer a realistic solution to help facilitate interactions between humans and machines—if only these agents were better equipped and more informed to hold up their end of an interaction. People and machines can interact to do things together, but in order to get the most out of every interaction, the agent must to be able to make reasonable judgements regarding your intent and goals for the interaction. We explore the concept of engagement from the different perspectives of the human and the agent. More specifically, we study how the agent perceives the engagement state of the other interactant, and how it generates its own representation of engaging behaviour. In this chapter, we discuss the different stages and components of engagement that have been suggested in the literature from the applied perspective of a case study of engagement for social robotics, as well as in the context of another study that was focused on gaze-related engagement with virtual characters.

L.J. Corrigan (✉) · G. Castellano
School of Electronic, Electrical and Systems Engineering,
University of Birmingham, Birmingham, UK
e-mail: ljc228@bham.ac.uk

C. Peters
Royal Institute of Technology (KTH), Stockholm, Sweden
e-mail: chpeters@kth.se

D. Küster
Jacobs University Bremen, Bremen, Germany
e-mail: d.kuester@jacobs-university.de

G. Castellano
Department of Information Technology, Uppsala University,
Uppsala, Sweden
e-mail: ginevra.castellano@it.uu.se

© Springer International Publishing Switzerland 2016
A. Esposito and L.C. Jain (eds.), *Toward Robotic Socially Believable
Behaving Systems - Volume I*, Intelligent Systems Reference Library 105,
DOI 10.1007/978-3-319-31056-5_4

4.1 Introduction

Human interactions are a product of millions of years of evolution, and as such they are typically smooth and effortlessly coordinated, benefiting massively from the fact that both interactants are able to draw upon a multitude of verbal and/or non-verbal cues in ways that help to regulate the interaction. If we are to interact with machines, in the way that the future depicts i.e., with robotic tutors, interactive display points, operator free terminals and the like, then we need to develop machines that can interact with us in a similarly intuitive fashion. However, at present, the way in which we interact with machines is strongly dictated by their design, which is often not optimal in terms of user experience, especially in cases where issues arise. Hence, our interest in building artificial agents, such as virtual characters and social robots with the ability to maintain interactions across a spectrum of task-orientated use cases has a strong applied perspective. Overall, there is growing interest in the engagement concept throughout the human-machine-interaction (HMI) and related fields, but what is engagement and why is it so important? In this chapter we hope to answer this question by unravelling this complex phenomenon, providing both new and existing HMI researchers with an firm underpinning of engagement related theory and concepts.

The remainder of this chapter is organised as follows: In the next section, we provide the reader with some general theory of the various stages and components of engagement. Additionally, we detail related concepts, such as the perception and generation of engagement related behaviours, and novel experimental considerations. In Sect. 4.3, we present two case studies: one of which looks at the perception of engagement for social robotics and another which considers the perception and generation of engagement related behaviours for virtual agents via gaze.

4.2 Theory

4.2.1 Fundamentals

When consulting a dictionary in the English language, the term engagement appears to be used in at least two different ways—as the starting or intention to start, referring to an initiation of contact, and again in the longer term sense, referring to engagement as something that is more involved. In the literature, engagement is defined in a number of ways: as a process; as a stage in a process, or the overall process; as an experience; as a cognitive state of mind; an empathic connection; or as a perceived or theorised indicator describing the overall state of an interaction. Nevertheless, there are two underlying fundamentals that are apparent across most engagement-related studies; the existence of various stages and components of engagement. In this section, we discuss each of these in turn.

4.2.1.1   Stages of Engagement

Engagement as a process can be analysed in terms of a number of discrete stages or phases. These may relate to the intensity or degree of involvement of a user with respect to the object or entity of engagement. For example, in a study of engagement with robots, Sidner and Dzikovska [29] refer to engagement as "a process by which individuals in an interaction start, maintain and end their perceived connection to one another". Most often, these stages are considered independently. For example, recognising the desire to *start* an interaction requires the system to detect an intention to engage, e.g., by tracking passers-by to ascertain if there are certain indicators which might suggest an initial interest to become involved with the system [22]. Whereas, to *maintain* an interaction suggests that the intention has been established and that the system must now adapt to the individual user in such a way that it keeps that user engaged for the term of the interaction. Failure to do so may cause the user to *end* the interaction before the system has achieved its purpose, for example to teach, inform or otherwise assist the user. To this extent, the system should be equipped to do both: detect when a user has irrecoverably ended an interaction, e.g., by getting up and walking away, whilst also being able to use appropriate behaviour to end an interaction once either party has achieved their purpose for becoming involved in the interaction.

It is a natural starting point to consider engagement as consisting of at least three broad stages, i.e., intention to engage, engaged and disengaged. However, O'Brien and Toms [17] refer to a fourth possible stage: re-engagement. The concept of re-engagement raises the important issue of when an interaction can be considered as complete. If either party is yet to achieve their purpose, but the user is showing signs of becoming disengaged, the system might try to utilise any information that is available, e.g., from current and previous interactions, in order to "understand" the underlying cause of the disengagement and then attempt to re-engage the user with a series of predefined strategies. However, in many cases, disengagement may be difficult to determine with certainty. For example, if the user looks away briefly, it may just mean that she has been temporarily distracted. In certain cases, looking away may in fact even signal engagement, such as during shared attention, when looking at an object under mutual consideration [21].

4.2.1.2   Components of Engagement

While engagement is frequently operationalized by means of measures of visual attention, it is important to distinguish conceptually between engagement and attention. Engagement is a complex phenomenon, a construct consisting of both cognitive (attention, concentration) and affective components (enjoyment) [17, 27].

Attention: As the cognitive component of engagement, attention is often characterised as a global on/off activity, whereas concentration is the ability to pay selective attention to one thing in particular, while ignoring others. For example, a user paying attention to a particular activity or object for a significant amount of time is concentrating. In our work, it is this form of selective attention (relating to concentration) that we are interested in, and in going forward we refer to this as just attention. This is therefore conceptually distinct from a global measure of wakefulness, or arousal, although the precise focus of selective attention may sometimes be more narrow, and sometimes be wider. Selective attention to a stimulus is a necessary component in most definitions in order for basic forms of engagement to occur. A more sustained form of attention provides a more elaborate requirement for engagement and also allows the possibility of affective involvement [25].

Another important factor of engagement is considering exactly what it is that a user is engaged with (i.e., the focus of engagement). This can generally only be inferred from the context, particularly for more sophisticated forms of engagement where there may be more than one potential focus of engagement. Gaze can signal attention [26], however, gazing at a particular object is not always indicative of attention. For example, the fact that a user is in the vicinity of a screen or is looking at one does not mean that they are paying attention to it (they may be day-dreaming for example), or that they are paying attention to those aspects that would be the most important ones from the perspective of the experimenter. In fact, even looking away from a screen does not allow the inverse inference that the subject has completely disengaged. Thus, while there is a certain probability that this is the case, looking away from the screen might simply indicate a moment in the interaction during which the user requires additional resources to process what was being said or presented. One way to improve confidence in assessing the attentional component of engagement in this situation is to consider only attention towards currently relevant aspects of the scene, in terms of gaze and other forms of attention related involvement and interaction. For example, in [20] during interaction with a virtual character, three qualities of engagement are defined, relating to the user (1) not looking at the screen at all, (2) looking at irrelevant aspects of the scene, and (3) looking at relevant aspects of the scene with respect to the ongoing interaction.

Enjoyment: As the affective component of engagement, enjoyment also plays a direct role in an interaction. For example, both positive and negative affect has been shown to influence student performance, motivation and effort [4]. More specifically, in terms of object focus, positive emotions such as enjoyment can increase the availability of cognitive resources, having a positive influence on the user's motivation, ability to utilise flexible learning strategies and self-regulation [18]. Positive affect also increases general motivation, which leads people to try harder in tasks, especially where they feel their effort will make a difference [11]. However, these emotions may not always be outwardly expressed, for example, a user is highly unlikely to smile or laugh throughout an interaction, nonetheless, enjoyment is an important component of the engagement construct. Here, enjoyment is most likely to be expressed indirectly, by continuing the interaction with a strong commitment to achieving certain goals. In gaming, for example, players who are immersed in a game tend to make very few facial expressions, but are nonetheless still very much enjoying the interaction. Here the effort afforded to the interaction could be associated with the positive affect (enjoyment), likewise the inverse could also be true, a lack of afforded effort could be associated with negative affect (boredom) [4].

4.2.2 Concepts

So far, in this chapter, we have only discussed engagement in terms of perception. However, if machines are to interact with humans in a natural and intuitive manner, it is not sufficient to focus entirely on one side of an interaction. Rather, we should consider engagement as a communicative process; a sender-receiver loop; to both perceive and generate engagement-related cues and signals. Here, we discuss the concept of engagement in terms of both perception and generation.

4.2.2.1 Perception

Perception refers to the use of the term engagement as it relates to the decoding of basic cues from another interactant, by a person or by a machine, for example by using computer vision techniques. Of general importance to our sense of engagement with others is our perception of their attention [10], which can be altered by factors such as the effect of distance between interactants on the salience of visual cues and the context of the situation, and by their enjoyment or at least our perception of their interest. Importantly, both cognitive and affective components of engagement can be measured (with a certain probability) on the basis of certain objective indicators and physiological measures. While this is not a one-to-one mapping between indicators and engagement, this allows a certain degree of automatic measurement of engagement that can be expected to become more reliable with the development of new sensors and algorithms. Enjoyment (pleasure) is the affective component of engagement that can be measured on the basis of several potential indicators, such as eyebrow activity (reverse sign) in combination with smiling. Here, eyebrow activity weighs more than smiling, and a moderately negative weight is added for lip-pressing and lip-tightening. Eyebrow movements can also be a predictor for concentration, obstacles and negative valence. Therefore, frowning may indicate effortful processing suggesting high levels of cognitive engagement which are likely to be associated with negative valence (depending on context and intensity). Smiling is expected to be a weak predictor of positive valence, but it might be effective for short-term social responsiveness [11]. Mouth movements, such as lip pressing and tightening can also be associated with task-related attention and concentration.

Furthermore, important non-verbal cues can be obtained based on head direction and gaze [1], blinking, eyebrow movement, posture and posture shifts [14], smiles [3], and engagement gestures [29]. These low-level signals can in some cases be interpreted as direct measures relating to engagement. However, typically, these individual measures become more informative when they are interpreted with respect to specific events in the task, or when they occur in a synchronized fashion with other indicators rather than individually.

4.2.2.2 Generation

The generation of cues and signals (i.e., their "encoding") requires at least an equal amount of attention. For example, during face-to-face interaction, the face generates a wealth of cues and signals that goes well beyond speech and facial expressions. It may be expected that we naturally pay attention to the face if we are engaged with that person, and may also display feedback such as nods to display our interest and/or show empathy by conducting appropriate facial expressions. In this respect, they might signal engagement, for example, by attending to the other and showing interest in what they say. An important distinction here, is whether such signals are based on a genuine interest or are superficial displays with the implicit or explicit purpose of communicating to the other that one is engaged. One may display signals of interest for a variety of superficial reasons, related to the accomplishment of high-level or abstract goals. Sometimes the display of interest is more important than the actual motivation [8]. In our previous work, an analysis of data extracted from explicit probes [5] and post-experiment questionnaires suggests that one's own perception of a robot, in terms of helpfulness, friendliness and attentiveness can help to maintain a type of engagement which lasts throughout an interaction [6]. An artificial agent capable of generating, or at least mimicking, certain engagement related behaviours could help to facilitate an intuitive interaction between humans and machines, giving the human the impression that the machine is intelligent enough to warrant further interaction.

4.2.3 Experimental Considerations

Engagement is often reduced to selective visual attention, perhaps as a practical consequence of a limited availability of measures within a given paradigm. However, in order to discuss engagement as a meaningful construct, we argue that the affective components have to be considered as well. Despite this, in practice, this can be difficult to achieve. As part of the work described in Sect. 4.3.1.2, we found the concept of annotating for the entire engagement construct to be extremely complex. In fact, we found ourselves asking "how should one annotate for both attention and enjoyment at the same time?". For this exact reason, we have started to explore engagement in a de-constructed format, considering engagement-related components individually.

4.2.3.1 Decomposition of the Engagement Construct

Attention and enjoyment are descriptive states in their own right. It is this decomposition of the engagement construct into cognitive and affective components that we believe can account for, and thus allow for, the fact that high engagement can, and often will, represent rather different socio-emotional states during an interaction. For example, a user may show evidence of intense cognitive engagement with a task, while the affective component might be anywhere between highly positive and highly negative. In the immediate situation, the assessment of attention and cognitive engagement with the task may initially be sufficient since it may not appear to matter how much a user is enjoying a task, as long as she/he continues to work hard to solve it. However, in order to anticipate eventual frustration and a high probability of disengagement in one of the subsequent tasks, it is essential that an engagement detector attempts to track also this affective component in order to facilitate appropriate and early interventions by the system.

4.2.3.2 Implicit Probes

During complex interactions, feeding sensor data directly into a computational model might not always be able to provide a accurate measure of the user's state of engagement. For example, in an educational interaction involving a robotic tutor, the system may need to understand why the user is showing signs of disengagement—is it that the task is too difficult, too easy, or is it because the user is simply discounting the advice provided by the robot? In this situation, the system could use a probe to answer some of these questions, i.e., by evaluating certain elements of the interaction.

So, what is a probe? A probe is a non-intrusive, pervasive method of extracting additional supporting features from within the interaction itself, providing highly standardised moments for analysis. Probes are pervasive in the sense that they can be integrated into any stage of an interaction [5]. They do not require the collection of any special additional types of data beyond the measures already stated in the consent forms. Rather, the probes define standardized situations that are naturally embedded into the flow of the task in such a way that they appear to the subject as a completely normal part of the interaction. Their standardization allows the formulation of substantially more meaningful predictions of behaviours within one experiment as well as between experiments and potentially even across different experimental paradigms. In this sense, they could also be described as modular building blocks that can be reused and which remain comparable even when other parts of the interaction require more flexibility. For example, experimenters in laboratory experiments, or doctors with a lot of experience in interviewing patients, will often use a similar approach using highly schematic questions and small talk in order to get a first sense of the participant or patient.

Furthermore, as probes only describe relatively short schematic modules with few degrees of freedom, they do not impede upon the natural flow of the interaction as opposed to, e.g., experimental designs that aim to obtain full control throughout the entirety of an experiment. For this reason, we argue that probes may be an ideal solution when needs for high levels of experimental control and analyses have to be balanced with maintaining a natural flow. Engagement, in this context, is a particularly relevant example because any measurement of engagement has to avoid disrupting the user engagement itself. The features extracted from these probes are embedded within the context of the main task, and are designed to provide the most accurate possible assessment of the user's engagement state. More specifically, we distinguish between two different types of probes: the social probe and the social-task probe. This is an additional step beyond the low-level continuous observation already employed elsewhere in HMI and related fields.

The design of a probe causes the user to respond in a certain way and it is that response which is then used to fortify the system's confidence that a user is in a particular state. For example, if the agent is unsure of the user's engagement state because confidence levels are low and social interaction hasn't occurred recently enough to make any inferences, then the agent can trigger a probe by attempting to socially engage the user in a one-to-one interaction. If the user stops what they are doing and responds to the agent's attempt, then we can increase the value associated with social engagement and also increase the overall confidence. Other metrics relating to immediacy, responsiveness and whether or not the user maintains their attention to the social interaction will further affect those values.

Social Probes: The social probe involves a simple, standardised piece of interaction between the agent and the user. Its purpose is to provide a standardised moment in which we can gauge how socially receptive the user is to the agent. To illustrate, in Fig. 4.1, we provide a time-line example of a social probe. The first three seconds are used to attract the attention of the user and the following two segments, lasting five seconds and three and a half seconds respectively, are used for analysis. As an example of this, if the user maintains gaze toward the agent across both maintainer segments, he/she is deemed as showing signs of high attention. We can also use this highly standardised piece of interaction to detect other non-verbal behaviours, such as smiles and facial expressions, including their temporal location within the probe.

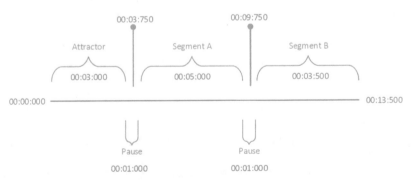

Fig. 4.1 An example of an interaction time-line for a social probe. Timings relate to the amount of interaction allocated to the attractor and two maintainer segments

Actual implemented examples of the content used in this particular type of probe are provided in Sect. 4.3.1.2.

Social-Task Probes: The social-task probe is concerned with the collaborative aspect of an interaction involving an agent and a user, such as the teacher-student or master-apprentice relationship. With this type of probe we can measure how receptive the user is to the agent's suggestions and assistance, e.g., by directing gaze toward specific items, or encouraging specific actions. From this we can also measure how reliant or independent the user is on assistance from the agent. In other words, social-task probes are designed to measure aspects of engagement in social-task interactions.

4.3   Practice

4.3.1   Case Study 1: Engagement in Social Robotics

To explore the engagement concept in a task-orientated scenario, we conducted a Wizard-of-Oz (WoZ) style data collection study using a robot. The study was carried out in the classroom environment of an English secondary school. The participants were children aged between 11 and 13; ten boys and ten girls. The demographics survey shows that all of the children had some experience using computers and knowledge of geography, but none had experience with robots. The WoZ-style approach was adopted as it is a common practice in HMI and related fields [24], allowing for a smoother and more believable interaction than what can be achieved on the basis of a fully autonomous robot in the early stages of development. With the help of the wizard, the robot can display realistic behaviours and respond to the child in a timely fashion, without having to implement a fully functioning autonomous system that was not yet available at the time of this case study.

4.3.1.1   Scenario

The children were asked, by their geography teacher, if they wanted to take part in an educational map reading activity with one-to-one support from a robotic tutor. However, the children were not informed, until after the study, that the robot was being controlled by the wizard, i.e., a human. The children were required to employ their existing geography-related knowledge, while also learning how to navigate the map using various combinations of the compass, ruler and map key. The robot provided the child with support that would not only help them to progress further in the task, but also help them to think about how the skills they were learning could be applied to a range of map-related problems. An activity script, using the appropriate level of difficulty, as identified in previous mock-up studies, was written and tested with the help of several teaching experts, ensuring that the content was in-line with the England and Wales National Curriculum for Geography. The robot,

Fig. 4.2 Child interacting
with the robot in a social
exchange

a NAO,[1] started each interaction by introducing himself and then asking the child
for his/her name, which was then repeated back to the child in a welcome statement.
Next, the robot provided the child with a brief tutorial, including an overview of
the activity to help familiarize the child with the interface, tools and the type of
support they could expect from the robot. The robot provided support throughout the
interaction, and at times when the robot did not need to intervene, it used several idling
animations to sustain a certain level of activity, realism, and presence. Additionally,
when addressing the child, the robot would attempt to maintain an acceptable level
of mutual gaze, looking away occasionally so not to be freaky. It was able to track
the child's face and maintain the gaze even as the child moved around in front of the
robot. The aim of these activities was to make the robot appear more intelligent and
lifelike.

4.3.1.2   Method

Technical Set-Up

The technical set-up for the WoZ study (*see* Fig. 4.2) comprises of a large touch-screen
table that was embedded horizontally into a supporting aluminium structure, forming
an interactive table-top surface, a torso-only version of the NAO humanoid robot,
three video cameras positioned in frontal, lateral and top-down locations, a Microsoft
Kinect,[2] an Affectiva Q Sensor[3] for measuring skin conductivity and OKAO vision
software by OMRON for measuring smile intensity and eye-gaze direction.

Implementation of Social Probes

Social Probe 1:
Attractor: *"Nice to meet you Joe"*
[PAUSE]
Maintainer (A): *"I hope that you are doing well today"*
[PAUSE]

---

[1]NAO, http://www.aldebaran-robotics.com/.

[2]Microsoft Kinect, http://www.microsoft.com/en-us/kinectforwindows/.

[3]Affectiva Q Sensor, http://www.qsensortech.com/.

Maintainer (B): *"and that you'll have fun hanging out with me for a little while"*

Social Probe 2:
Attractor: *"Joe, Have you ever seen Wallace and Grommit?"*
[PAUSE]
Maintainer (A): *"I think that there is actually a Wallace and Gromit film with robotic trousers."*
[PAUSE]
Maintainer (B): *"I wonder what it would be like to have legs myself"*

Social Probe 3:
Attractor: *"Thank you so much for all you help!"*
[PAUSE]
Maintainer (A): *"I hope you had a good time!"*
[PAUSE]
Maintainer (B): *"I thought you did really good!"*

Level of Automation

All aspects of the interaction, including robot control, helping the child to progress through the activity, implementing the correct teaching strategies, and engaging in social exchanges, were remotely controlled by a qualified teacher using a bespoke interaction control interface (see Fig. 4.3).

Data Collection

The corpus of data collected from *case study 1* included more than seven hours of video material for each of the three viewing angles, interaction data recorded from involvement with the activity and data from the low-level sensors, such as skin conductance, facial action units, smile intensity, posture-related lean information and gaze direction. Data was cleaned of certain artefacts, including sensor noise and incomplete cases, and the raw low-level information was binned into discrete instances of time, more specifically 250 ms, which allowed us to process, analyse and model the data using statistical and machine learning methods.

Fig. 4.3 Wizard's view of the interaction control interface

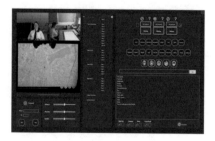

Annotating the Video Material

The simplest method of annotating video material is to use relative ratings and discrete segments of video media. However, in this work we require continuous measures that allow at least a rough estimation of the timing of relevant changes. Such continuous annotation data has the potential to add substantial flexibility to use the final computed 'ground truth' for different purposes, such as statistical analysis and training of machine learning algorithms. However, the trade-off for obtaining continuous data is that the precise moment of changes in subjective states such as engagement can be difficult to pinpoint even for trained raters, resulting in an overall lower reliability compared to a single global Likert scale. Nevertheless, this work follows the more novel approach of adopting the continuous measure which could, potentially, encode far more interesting information.

Annotation Software: Off-the-shelf annotation software could not provide the flexibility to perform continuous annotations that allow the simultaneous presentation of multiple video streams of data, and the input modality is typically fixed to either mouse or keyboard. We wanted to explore the use of a game-pad, or more specifically the thumb stick of a game-pad. The assumption here is that releasing the thumb stick can be used naturally to indicate a return of the annotated measure back to a neutral state extremely quickly, whereas a mouse or keyboard would cause periods of uncertainty in the output signal, due to the fact that some active effort and time is required to return the rating back to neutral. Furthermore, the latter modalities are unable to offer the same fine grained resolution as an analogue thumb stick. CAT, or Continuous Annotation Tool, is a custom solution designed to facilitate these seemingly "unusual" annotation requirements, i.e., synchronously displaying three different views of the interaction such as, e.g., frontal, lateral and top-down, and providing a simple visual representation of the rating intensity (in real-time) on a vertical slider bar, which transitions from green at the very top to represent a positive or high intensity, orange in the centre to represent a neutral intensity and red at the bottom to represent an extremely low intensity (see Fig. 4.4). In CAT, annotator ratings are automatically logged with two decimal point precision, with maximum and minimum extremes set to 1 and -1 respectively.

Fig. 4.4 CAT: Continuous Annotation Tool: Bespoke software, developed specifically for the use with continuous ratings

Annotators: For each annotated signal, i.e., social attention and valence, we used the same three annotators. So, for social attention, the annotators were: (1) a pedagogical researcher who could look at attention from a teacher-student perspective, (2) a psychologist to look at attention from a behavioural aspect, and (3) a researcher, specialising in automatic non-verbal behaviour analysis for social robotics.

Annotator Agreement: The issue of reaching agreement is an important part of the annotation process. Our methodology was to reach an acceptable level of agreement in advance, ensuring that the final output signals were the best that we could achieve with the time and resources we had. We adopted a three-step approach, starting with a discussion of the overall objective criteria, in an attempt to pre-emptively list potential indicators, and later in the process the annotators were asked to produce a voice over account and a single continuous rating for the same randomly selected interaction, to visualise the different output signals in a side-by-side analysis. Obviously, there were differences between the signals, but an acceptable level of difference, i.e., less than a second, was achieved.

Ground Truth Extraction

Computing and then extracting a ground truth is an essential process for this type of non-verbal behaviour recognition. The output of the extraction process, which involves aggregating the ratings from multiple annotations into a single signal, represents the final measure for a particular criterion, such as social attention or valence. Producing a ground truth can be a relatively straightforward process when working with discrete labels, but the very nature of our continuous rating process renders many existing methodologies as impractical [13]. In fact, many researchers choose to completely ignore the concept of agreement in favour of simpler methods, such as using the mean from several ratings, or alternatively opting to manually assess the ratings [16]. We wanted to ensure that we were not introducing biases or losing information, so we opted to explore other more suitable methods.

An in-depth review of the literature uncovered two potential methods for computing a ground truth, based on annotator agreement. The first method focuses on the use of a correlative threshold, specifically 0.45, meaning that ratings from annotator pairs with correlative coefficients smaller than 0.45 are quite simply omitted from the computation of the ground truth. In contrast to this method, we consider it to be of the utmost importance that the ratings from *all* annotators are included when computing the ground truth, even those who are in disagreement with the others. This provides a more realistic ground truth that takes into account the different backgrounds and perspectives of the annotators. Therefore, we adopted the alternative method of using weighted correlations, similar to the work by Nicolaou et al. [16], which we then further extended for our interaction length non-segmented continuous signals. Here, ratings from all annotators are considered in the computation with the condition that the most highly correlated annotator pairs are given more weight than disagreeing annotators (see Fig. 4.5 for an example of the output ground truth signal). The graph in Fig. 4.6 provides an estimate of inter-rater reliability, in terms of intraclass correlation (ICC), more specifically, we have used ICC (2,3), which denotes the ICC

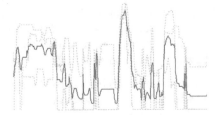

Fig. 4.5 Segment showing the continuous measure of social attention from three annotators (*light dashed lines*) with the final computed 'ground truth' (*dark solid line*)

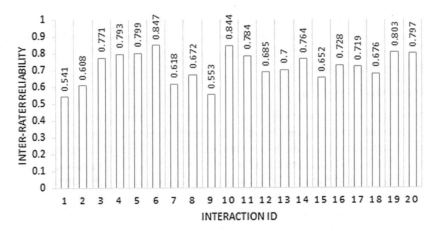

Fig. 4.6 Graph showing an estimate of inter-rater reliability for each of the twenty child-robot interactions, i.e., in terms of intraclass correlation (ICC)

values are calculated for each interaction using Case 2 from the work by Shrout and Fleiss [28], involving the same three annotators for each case.

4.3.1.3 Analysis and Results

Analysis of the corpus has, so far, been two-fold: an *interaction-length* analysis of the gaze with the social attention signal, and a *social probe-interval* focussed analysis of the behaviour-related variables with the social attention and valence signals. The motivation for this analysis is to explore what features may be descriptive of engagement in HMI.

Analysis A: Interaction-length

Here, we report the results of a point biserial correlation analysis between the social attention signal (interval scale) and a pre-processed dichotomous nominal scale relating to robot gaze, the two levels are 0 (if the learner was not looking at the robot)

Table 4.1 Results of a point biserial correlation analysis between social-related attention and gazing at the robot

ID	r_{pb}	Sig.	N	ID	r_{pb}	Sig.	N
1	0.49	$p < 0.001$	3234	11	0.57	$p < 0.001$	2290
2	0.41	$p < 0.001$	4124	12	0.25	$p < 0.001$	2028
3	0.4	$p < 0.001$	1822	13	0.4	$p < 0.001$	2258
4	0.55	$p < 0.001$	1650	14	0.41	$p < 0.001$	2160
5	0.59	$p < 0.001$	2744	15	0.45	$p < 0.001$	2024
6	0.64	$p < 0.001$	2184	16	0.42	$p < 0.001$	2362
7	0.45	$p < 0.001$	1954	17	0.39	$p < 0.001$	2214
8	0.47	$p < 0.001$	2086	18	0.45	$p < 0.001$	1874
9	0.62	$p < 0.001$	2258	19	0.56	$p < 0.001$	1566
10	0.38	$p < 0.001$	1826	20	0.51	$p < 0.001$	1586

The columns represent interaction ID, point biserial correlation coefficient, statistical significance and number of low-level instances used in the analysis, respectively

and 1 (if the learner is looking at the robot). The results of the analysis, set out in Table 4.1, show that, on average, gazing at the robot is moderately correlated with social attention ($r_{pb} = 0.47$, $p < 0.01$).

Analysis B: Social Probe-interval

To understand which, if any, behaviour-related variables are an indication of engagement, samples extracted from the *social probes* are compared with samples taken from similar areas of the interaction, +/-30 s, i.e., samples extracted from areas that do not involve *social probes*, in this work we will refer to these as samples taken from *"non-social probes"*. The samples taken from the *social probes* have been extracted in a way that they surround and capture the entire piece of probe-related interaction. There are three *social probes* embedded into each interaction and the duration of each probe was: 17.75, 13.5, and 6.25 s, respectively, with no overlap between segments (*see* Sect. 4.3.1.2). The *non-social probe* samples were extracted from other random moments outside of the probe-intervals, using an identical process. This process was repeated for each of the 20 interactions, providing a total of 2420 instances of raw low-level data for each case. A comparison of the *social probes* with just a single case of *non-social probes* does not actually tell us anything useful, therefore, we extracted three different cases of *non-social probes* to further support our analysis.

Social Attention

For social attention we consider information from gaze, smile and facial expressions. Pearson product-moment correlation coefficients have been computed to assess relationships between gaze, smile and facial expressions, and the social attention signal, obtained from the ground truth extraction process (Sect. 4.3.1.2). The results of the analysis are set out in Table 4.2.

Table 4.2 Results of a Pearson product-moment correlation between the behavioural indicators and the social attention signal, i.e., for samples taken from the *social probes* and three other *non-social probes* (NSP)

	Social attention signal			
	Social Probe	NSP 1	NSP 2	NSP 3
Smile	0.329**	0.325**	0.359**	−0.082**
AU1 inner brow raise	0.046*	0.085**		0.052*
AU2 outer brow raise	−0.102**	−0.157**	−0.073**	
AU4 brow lower	0.041*	0.094**	0.046*	0.067**
AU5 eye widen				0.237**
AU9 nose wrinkle	0.187**	0.241**	0.308**	0.160**
AU10 lip raise		0.083**		0.187**
AU12 lip corner pull		−0.222**	−0.116**	−0.178**
AU14 dimpler	0.137**			
AU15 lip corner dep	−0.169**			
AU17 chin raise	−0.218**		−0.040*	−0.077**
AU20 lip stretch	−0.197**	0.134**		0.107**
AU6 cheek raise	0.134**	0.219**	0.200**	0.245**
AU7 lids tight	0.152**	−0.115**	−0.100**	−0.062**
AU18 lip pucker		−0.146**	−0.153**	
AU23 lip tightener	−0.111**	−0.163**	−0.281**	−0.170**
AU24 lip presser	−0.183**	0.094**		0.068**
AU25 lips part				−0.060**
AU26 jaw drop	0.104**	0.291**	0.189**	0.143**
AU28 lips suck	0.160**			−0.086**
AU45 blink eye closure	0.054**	0.051*		
NormEDAValue	−0.230**	−0.218**	−0.150**	−0.065**
Gaze@Robot	0.498**	0.537**	0.456**	0.446**
Gaze@Table L	−0.281**	−0.134**	−0.070**	
Gaze@Table R	−0.331**	−0.294**	−0.281**	−0.340**
Gaze@Elsewhere				−0.041*

This table only shows significant correlations, figures marked with ** represent significance at the 0.01 level and others marked with * represent significance at the 0.05 level

Valence

For valence, the affective component of engagement, we focus on behavioural indicators. Pearson product-moment correlation coefficients have been computed to assess relationships between the smile and facial expressions, and the valence signal, obtained from the ground truth extraction process (Sect. 4.3.1.2). The results of the analysis are set out in Table 4.3.

Table 4.3 Results of a Pearson product-moment correlation between the behavioural indicators and the valence signal, i.e., for samples taken from the *social probes* and three other *non-social probes* (NSP)

	Valence signal			
	Social probe	NSP 1	NSP 2	NSP 3
Smile	0.524**	0.532**	0.537**	0.108**
AU1 inner brow raise	0.055**	0.069**		0.061**
AU2 outer brow raise	−0.077**	−0.100**	−0.071**	−0.072**
AU4 brow lower	−0.099**	−0.066**	−0.079**	−0.086**
AU5 eye widen				0.098**
AU9 nose wrinkle			0.125**	
AU10 lip raise	−0.118**			
AU12 lip corner pull	0.179**	−0.075**		−0.043*
AU14 dimpler	0.320**	0.171**	0.154**	0.135**
AU15 lip corner dep		0.180**	0.150**	
AU17 chin raise				−0.059**
AU20 lip stretch	−0.130**	0.124**	0.122**	0.068**
AU6 cheek raise	0.210**	0.055**	0.089**	0.067**
AU7 lids tight	0.429**	0.158**	0.132**	0.067**
AU18 lip pucker	0.205**	0.069**		0.102**
AU23 lip tightener	−0.227**	−0.232**	−0.241**	−0.129**
AU24 lip presser	−0.230**	0.078**	0.067**	0.054**
AU25 lips part	−0.075**		0.089**	
AU26 jaw drop	0.067**	0.089**	0.106**	0.054**
AU28 lips suck	0.152**	0.051*	0.070**	
AU45 blink eye closure			0.079**	

This table only shows significant correlations, figures marked with ** represent significance at the 0.01 level and others marked with * represent significance at the 0.05 level

4.3.1.4 Discussion

The most obvious finding to emerge from the analysis of the *social probe* versus *non-social probe* samples, is that correlations appear to be stronger, in the majority of cases, in samples taken from the *social probes*. A possible reason for this is that the social probes can be expected to have been particularly engaging in the sense of a simultaneous recruitment of different highly over-learned behavioural response systems. That is, in this case, the existence of well learned norms appears to have led more clearly and consistently communicated social signals—whereas, in the *non-social probe* case there are substantially less contextually defined social schemata to help guide the encoding as well as decoding of the behaviours.

4.3.2 Case Study 2: Engagement with Virtual Agents

The development of autonomous virtual agents and animated characters capable of engaging humans in real-time interaction faces many of the same challenges as similar attempts using physical embodiments such as social robots. These include the task of obtaining robust real-time detection of engagement-related behaviour from human users, timely responses from virtual agents, and the generation of appropriate behaviours by agents that are capable of properly expressing their state of engagement and focus of attention. This section describes an example scenario that involved engagement between a human and a virtual agent. The interaction was primarily shaped by gaze behaviour, in particular when it was directed at predefined objects and locations within an artificial scene (for a review, see [26]).

4.3.2.1 Gaze Detection and Representation

There are, in principle, a large number of measures that can be used to detect user engagement. These include, for example, monitoring verbal and non-verbal behaviours, taking physiological measurements and tracking task related actions that are conducted inside an application or virtual environment. However, not all of these measured may be needed at once for a basic analysis of engagement processes. In this scenario, the main method for detecting engagement was based on the gaze behaviours of the user as they engaged in an object identification task with a virtual agent capable of referring to objects non-verbally through its own gaze behaviours.

Gaze Detection

The gaze detection system used facial feature analysis of the image captured from a standard web camera to capture the user's gaze direction (head and eye directions) inside and outside of the screen. The detection process commenced with the eye-centers, which are easily detected, in order to allow the estimation of the eye corners and eyelids and positions on the eyebrows, nostrils, and mouth region. These were subsequently tracked using a Lucas Kanade tracker [12], capable of operating under a wide variety of conditions. Head-pose estimation was calculated based on the displacement of the midpoint of the eye centers from an initial head position in which the user was facing the screen frontally.

Embedded Representations

Since the user's focus of attention is usually highly transient as it shifts around a scene, it can be informative to use predefined objects to track these changes more systematically. Virtual Attention Objects (VAOs) simplify the analysis of what is being looked at in the scene by storing, on a per-object level, when and how much each part of the scene has been looked at. A single VAO is attached to each scene object for which we wish to accumulate user attention information. For example, a single VAO may be defined for each visible object in the scene, including the virtual agent itself. Depending on the requirements of the application, the virtual agent may

Fig. 4.7 A scenario involving a virtual agent and several objects. The user's gaze behaviour (*left*) is tracked as the agent conducts gaze behaviours towards various objects (*centre*). Gaze and attentive behaviours towards specific elements of the scene are recorded in real-time through Virtual Attention Objects (*right*) as a basis for monitoring user engagement in the scenario [21]

be represented by a single VAO, or a separate VAO may be defined for each part of the virtual agent for which information about user attention is required. VAOs may also be defined for more abstract objects. For example, a single VAO may be defined to represent the area outside of the screen. This VAO can then be used to record whenever the user gaze wanders outside of the scenario area, which renders it a useful metric for disengagement.

Furthermore, VAOs operate in a simple manner: Screen coordinates relating to user gaze are resolved to the specific associated VAO or VAOs. On this basis, the estimated level of attention can be adapted accordingly. The combined VAOs therefore represent a history of how much and when the user has fixated on each object in the scene. Figure 4.7 illustrates a virtual scene from [21], including a virtual agent and a number of objects, and an accompanying VAO representation.

4.3.2.2   Engagement Modelling

In this work, the focus was not only on the different components of engagement, but also the level and quality of those components. Expanding measures beyond what has already been discussed in this chapter, in order to capture varying degrees of engagement and related components. More specifically, we refer to directedness, level of attention, level of engagement and quality of engagement.

Directedness relates to the momentary orienting of the user's body parts with respect to another entity or object from the perspective of that entity or object. The metric is inspired by Baron-Cohen's eye, head and body direction detectors [2] and related work [21]. Directedness as a concept refers to transient and momentary processes that alone do not imply attention or engagement. For example, high directedness was assumed if the eye and head direction was sampled from a user while they are in the process of a gaze change to an alternative location. However, the consideration of directedness over time (and with respect to other aspects of the scene and user) is an important building block towards a model of engagement measurement. Depending on the scenario and, especially, the sensory capabilities of the detection system, the directions of the eyes, head, body and even locomotion trajectories may contribute to directedness measurements.

Level of attention is based on directedness and refers to gaze falling within certain regions over a period of time. It therefore corresponds to the concept of a focus of attention and corresponding dwell time by the eye. An important issue in this respect relates to the clustering of the foci of interest of the user. For scenes that are clearly composed of objects, VAOs may be used as one method for clustering fixation locations on a per-object basis.

Level of interest is based on the stored attention levels over time for each member of a set of VAOs. Each member is categorised according to whether it is a scene object, the agent, the background, or a special object representing the area outside of the screen. It is at this level that specific forms of context can be accounted for: By dynamically defining a set of VAOs containing only those objects relevant to the current interaction, such as recently pointed to or discussed objects, the attention of the user can be compared with this set to obtain a measurement of their level of interest in the interaction itself, referred to here as the level of engagement.

Level of engagement encapsulates how much the user has been looking at the relevant objects in the scene at appropriate times. These will be recently referenced objects in the interaction, e.g. those looked at, pointed to and/or verbally described. These measures are made possible by considering the specific set of VAOs corresponding to currently and recently referenced objects in the interaction. When the agent is talking, but does not refer to anything in the environment, it will be the only VAO in the set, and when it stops talking, this VAO set will be empty.

Quality of engagement accounts for the fact that attention paid to the scene does not necessarily indicate engagement in relation to a particular activity, in this case, interaction with the agent. For example, the user may be looking at the scene for superficial reasons without engaging in the interaction. It provides a slightly more detailed assessment of the type of engagement that the user has entered into. For example, a user who is not engaged in the interaction may not necessarily be looking outside of the scene. Instead, they may be attending to the scene in a superficial manner, looking at objects of interest that are irrelevant to the ongoing interaction. In particular, they may appear to be affectively disengaged from the interaction. We therefore define three broad quality levels: (i) engaged in the interaction (ii) superficially engaged with the scene and action space and (iii) uninterested in the scene/action space. In this way, the behaviour of the user is not being considered in isolation, but in the context of what the agent is doing. If the agent is describing something important for example, a user's disengagement can be considered more serious than if the agent is not doing anything at all.

4.3.2.3 Human Perception

In order to support sustained interactions between humans and virtual agents, models of engagement should not only account for the detection of behaviour from users and the generation of expressive behaviours by agents, but also how humans perceive artificially generated behaviours. This is an important consideration that may involve many factors, such as the embodiment and expressive capabilities of the system, the qualities of the expressive motion, the effects of context on perception and more. One of the fundamental methods of signalling attention and engagement by artificial systems involves gaze [26] and numerous studies have considered the human perception of gaze and other attention related behaviours made by artificial systems.

Gaze and Direction of Attention Perception

Gaze direction perception, which primarily involves eye and head movements and their relationships with objects in the environment from the perspective of a witness [23], can be extended to a more general concept that relates to the perception of the direction of attention of others. Here, the orientation of the eyes, head, body, and even locomotion trajectories, may contribute to one's impression that they are potentially being attended to by others. For example, at a distance, the eyes of the other may not be clearly visible in which case head and body directions may offer more prominent clues as to the direction that the other is attending to.

Opening Interactions

An example of the use of the concept of direction of attention perception is proposed in [19] for the purposes of opening interactions between interactants in virtual environments and the human perception of the attentive behaviours of virtual characters. Kendon [9] describes a sequence of opening cues for meeting interaction, noting interaction fundamentals, such as the requirements for participants to see each other, while Goffman notes [7] that we generally seek to avoid the social embarrassment of engaging in interaction with an unwilling participant. Therefore, some degree of confidence of interaction reciprocation is necessary, through subtle gaze behaviours for example, before more explicit actions are made, such as verbal greetings. Based on this, a model of engagement opening is described for virtual agents based on a combination of their own gaze behaviours towards others that they wish to interact with and also their interpretations of the gaze and locomotion behaviours of others.

4.3.2.4 Discussion

Important future work is waiting to be done to establish better relationships between the expressive qualities of behaviours of virtual characters, their style and embodiments, and human perception. In humanoid characters, for example, small movements, such as saccadic eye movements and eye-blinks, may play important roles in signalling that the character is concentrating heavily, rather than simply daydreaming or momentarily unresponsive. Embodiment issues related to the use of virtual characters versus their physical counterparts is also an interesting research area. For example, [15] have found that faces that are physically-projected onto face-like surfaces may have a number of advantages over virtual faces constrained to flat screens, especially in relation to the character's ability to engage with individual users and for cueing real objects in the environment. The studies represent some important foundations for constructing artificial entities capable of engaging humans.

Acknowledgments This work was partially supported by the European Commission (EC) and was funded by the EU FP7 ICT-317923 project EMOTE (EMbOdied-perceptive Tutors for Empathy-based learning) and the EU Horizon 2020 ICT-644204 project ProsocialLearn. The authors are solely responsible for the content of this publication. It does not represent the opinion of the EC, and the EC is not responsible for any use that might be made of data appearing therein.

References

1. Asteriadis S, Karpouzis K, Kollias S (2009) Feature extraction and selection for inferring user engagement in an hci environment. In: Human-computer interaction. Springer, New Trends, pp 22–29
2. Baron-Cohen S (1994) How to build a baby that can read minds: cognitive mechanisms in mind reading. Curr Psychol Cogn 13:513–552
3. Castellano G, Pereira A, Leite I, Paiva A, Mcowan PW (2009) Detecting user engagement with a robot companion using task and social interaction-based features interaction scenario. In: Proceedings of the 2009 international conference on multimodal interfaces, pp 119–125
4. Christenson SL, Reschly AL, Wylie C (2012) Handbook of research on student engagement. Springer, Boston
5. Corrigan LJ, Basedow C, Küster D, Kappas A, Peters C, Castellano G (2014) Mixing implicit and explicit probes: finding a ground truth for engagement in social human-robot interactions. In: Proceedings of the 2014 ACM/IEEE international conference on HRI. ACM, pp 140–141
6. Corrigan LJ, Basedow C, Küster D, Kappas A, Peters C, Castellano G (2015) Perception matters! Engagement in task orientated social robotics. In: IEEE RO-MAN 2015. doi:10.1109/ROMAN.2015.7333665
7. Goffman E (2008) Behavior in public places. Simon and Schuster, New York
8. Kappas A, Krämer N (2011) Studies in emotion and social interaction. Face-to-face communication over the internet: emotions in a web of culture, language, and technology. Cambridge University Press, Cambridge
9. Kendon A (1990) Conducting interaction: patterns of behavior in focused encounters, vol 7. CUP Archive, Cambridge
10. Langton SR, Watt RJ, Bruce V (2000) Do the eyes have it? Cues to the direction of social attention. Trends Cogn Sci 4(2):50–59
11. Lewis M, Haviland-Jones JM, Barrett LF (2010) Handbook of emotions. Guilford Press, New York
12. Lucas BD, Kanade T (1981) An iterative image registration technique with an application to stereo vision. In: Proceedings of the 7th international joint Conference on Artificial Intelligence, pp 674–679
13. Metallinou A, Narayanan S (2013) Annotation and processing of continuous emotional attributes: challenges and opportunities. In: 2013 10th IEEE international conference and workshops on automatic face and gesture recognition (FG), pp 1–8
14. Mota S, Picard RW (2003) Automated posture analysis for detecting learner's interest level. In: Conference on computer vision and pattern recognition workshop, 2003. CVPRW'03, vol 5. IEEE, pp 49–49

15. Moubayed SA, Edlund J, Beskow J (2012) Taming Mona Lisa: communicating gaze faithfully in 2d and 3d facial projections. ACM Trans Interact Intell Syst (TiiS) 1(2):11
16. Nicolaou MA, Gunes H, Pantic M (2010) Automatic segmentation of spontaneous data using dimensional labels from multiple coders. In: Proceedings of LREC int'l workshop on multimodal corpora: advances in capturing, coding and analyzing multimodality, pp 43–48
17. O'Brien HL, Toms EG (2008) What is user engagement? A conceptual framework for defining user engagement with technology. J Am Soc Inf Sci Technol 59(6):938–955
18. Pekrun R, Elliot AJ, Maier MA (2009) Achievement goals and achievement emotions: testing a model of their joint relations with academic performance. J Educ Psychol 101(1):115
19. Peters C (2006) Evaluating perception of interaction initiation in virtual environments using humanoid agents. In: Proceedings of the 2006 conference on ECAI 2006: 17th European conference on artificial intelligence 29 Aug–1 Sept, 2006. IOS Press, Riva del Garda, pp 46–50
20. Peters C, Asteriadis S, Karpouzis K, de Sevin E (2008) Towards a real-time gaze-based shared attention for a virtual agent. In: Workshop on affective interaction in natural environments (AFFINE), ACM international conference on multimodal interfaces (ICMI08)
21. Peters C, Asteriadis S, Karpouzis K (2010) Investigating shared attention with a virtual agent using a gaze-based interface. J Multimodal User Interfaces 3:119–130. doi:10.1007/s12193-009-0029-1
22. Pitsch K, Kuzuoka H, Suzuki Y, Sussenbach L, Luff P, Heath C (2009) The first five seconds: contingent stepwise entry into an interaction as a means to secure sustained engagement in hri. In: The 18th IEEE international symposium on robot and human interactive communication, 2009. RO-MAN 2009, pp 985–991. doi:10.1109/ROMAN.2009.5326167
23. Qureshi A, Peters C, Apperly I (2013) Interaction and engagement between an agent and participant in an on-line communication paradigm as mediated by gaze direction. In: Proceedings of the 2013 inputs-outputs conference: on engagement in HCI and performance, p 8
24. Riek LD (2012) Wizard of oz studies in hri: a systematic review and new reporting guidelines. J Hum-Robot Interact 1(1):119–136
25. Roseman IJ, Smith CA (2001) Appraisal theory: overview, assumptions, varieties, controversies. In: Appraisal processes in emotion: theory, methods, research. Series in affective science, pp 3–19
26. Ruhland K, Andrist S, Badler J, Peters C, Badler N, Gleicher M, Mutlu B, Mcdonnell R (2014) Look me in the eyes: a survey of eye and gaze animation for virtual agents and artificial systems. In: Eurographics state-of-the-art report. The Eurographics Association, pp 69–91
27. Shernoff DJ, Csikszentmihalyi M, Schneider B, Shernoff ES (2003) Student engagement in high school classrooms from the perspective of flow theory. Sch Psychol Q 18(2):158–176
28. Shrout PE, Fleiss JL (1979) Intraclass correlations: uses in assessing rater reliability. Psychol Bull 86(2):420–428
29. Sidner CL, Dzikovska M (2005) A first experiment in engagement for human-robot interaction in hosting activities. Advances in natural multimodal dialogue systems, pp 55–76

Chapter 5
Social Development of Artificial Cognition

Tony Belpaeme, Samantha Adams, Joachim de Greeff,
Alessandro di Nuovo, Anthony Morse and Angelo Cangelosi

Abstract Recent years have seen a growing interest in applying insights from developmental psychology to build artificial intelligence and robotic systems. This endeavour, called developmental robotics, not only is a novel method of creating artificially intelligent systems, but also offers a new perspective on the development of human cognition. While once cognition was thought to be the product of the embodied brain, we now know that natural and artificial cognition results from the interplay between an adaptive brain, a growing body, the physical environment and a responsive social environment. This chapter gives three examples of how humanoid robots are used to unveil aspects of development, and how we can use development and learning to build better robots. We focus on the domains of word-meaning acquisition, abstract concept acquisition and number acquisition, and show that cognition needs embodiment and a social environment to develop. In addition, we argue that Spiking Neural Networks offer great potential for the implementation of artificial cognition on robots.

Keywords Human-Robot interaction · Social robots · Cognitive systems · Spiking artificial neural networks

5.1 Introduction

The recent, but fast expanding technological and financial investments in the production of intelligent robots rely on the design of robots with *effective and believable sensorimotor, cognitive and social capabilities*. For example, robots acting as assistive and social companions for the elderly must be able to autonomously navigate in the private home (or care home) where the elderly person lives, have fine manipulation skills to handle objects, be capable of understanding and using natural language for communication, and have believable social skills to enrich the experience of

T. Belpaeme (✉) · S. Adams · J. de Greeff · A. di Nuovo · A. Morse · A. Cangelosi
Plymouth University, Centre for Robotics and Neural Systems, Plymouth, UK
e-mail: tony.belpaeme@plymouth.ac.uk

© Springer International Publishing Switzerland 2016
A. Esposito and L.C. Jain (eds.), *Toward Robotic Socially Believable
Behaving Systems - Volume I*, Intelligent Systems Reference Library 105,
DOI 10.1007/978-3-319-31056-5_5

its elderly owner. Moreover, robots must be able to adapt to the requirements of the specific user, to react dynamically to changing environments and to learn new behavioural and cognitive skills through social interaction with the human user.

Cognitive robotics offers a feasible methodology for the design of robots with adaptive and learning capabilities, which can develop new skills through social interaction and learning. Cognitive robotics is a subfield of robotics in which robots are built based on insights gleaned from psychology, physiology and neuroscience, with the goal of replicating human-like performance on artificial systems [20, 78]. Cognitive robots are—as opposed to industrial robots—intended to work in open, unstructured and dynamic environments, the environments in which people typically feel at home, but robots do not. If someone asks a child to give a cup of water, the child can recognise and grasp the intended cup from among other objects, offer it, and do that while having a conversation. All this seems effortless to the child, but robots are—at this time—not able to do this in an open and dynamic environment. Robots might be programmed or trained to hand over a cup in a carefully controlled environment, but this would not generalise to handing over, say, a towel. As a rule of thumb, anything that seems effortless to humans is currently very hard for robots. And vice versa, we can build artificially intelligent systems and robots that can do things—such as playing chess or welding at a precise rate—that only very few of us ever master.

So why do classical approaches to building artificial intelligence and robots, that serve well to build chess playing computers and plan assembly tasks, fail to build AI that deals with unstructured and dynamic problems? The answer might lie in the study of development: young children seemingly effortlessly pick up skills which are very hard or impossible for robots to master. The question presents itself: can the same processes that are so successful in growing children be used to build intelligent robots? *Developmental robotics* is the interdisciplinary approach to the autonomous design of a complex repertoire of sensorimotor and mental capabilities in robots that takes direct inspiration from the developmental principles and mechanisms observed in the natural cognitive systems of children [7, 16, 79]. Developmental robotics relies on a highly interdisciplinary effort of empirical developmental sciences such as developmental psychology and neuroscience, and computational and engineering disciplines such as robotics and artificial intelligence. Developmental sciences provide the empirical bases and data to identify the general developmental and learning principles, mechanisms, models, and phenomena guiding the incremental acquisition of cognitive skills. The implementation of these principles and mechanisms into a robots control architecture and the testing through experiments where the robot interacts with its physical and social environment simultaneously permits the validation of such principles and the actual design of an increasingly complex set of complex behavioural and mental capabilities in robots.

Developmental robotics follows a series of general principles that characterise its approach to the design of intelligent behaviour in robots. Two of the key principles are the exploitation of embodiment factors in the development of cognitive capabilities and the focus on social learning.

Embodiment concerns the fundamental role of the body in cognition and intelligence. As Pfeifer and Scheier [57] claim, "intelligence cannot merely exist in the form of an abstract algorithm but requires a physical instantiation, a body" (p. 694). In psychology and cognitive science, the field of embodied cognition (also known as grounded cognition [10]) demonstrates the important roles of action, perception, and emotions in the grounding of cognitive functions such as memory and language [55]. For example, sensorimotor strategies, as postural changes, support the child in the early acquisition of words [63]. Gestures like pointing and finger counting are crucial in the acquisition of number knowledge [3]. Such developmental psychology studies are consistent with neuroscience evidence on embodied cognition, as brain-imaging studies showing that higher-order functions such as language share neural substrates normally associated with action processing [59]. The principle of social learning in developmental psychology is based on child development research on the role of social learning capabilities (instincts) in the very first days of life. This is evidenced for example by observations that newborn babies have an instinct to imitate the behavior of others and can imitate complex facial expressions after just few hours from birth [51]. Moreover, comparative psychology studies have shown that 18–24-month-old children have a drive to cooperate altruistically, a capacity missing in our closest genetic relatives as chimpanzees [80].

This chapter offers a summary of two recent studies on the modelling of embodiment and social learning in developmental humanoid robots. Both use the iCub humanoid platform both to exploit the properties of humanoid body configuration for embodiment modelling purposes and also for the benefits of using such humanoid platforms in social robotics scenarios. We will also discuss the potential of neuromorphic methods and hardware, as a first step for a brain-inspired approach to modelling the embodied basis of cognitive and communicative skills.

5.2 Why Embodiment Matters

Embodiment matters, not only in the development of natural cognition, but also in constructing *artificial* cognition. The brain or, in the case of robots, the control software cannot be seen as separate from the body in which it operates. Human cognition is deeply rooted in the shape of our bodies and how our bodies interact with the world. Likewise, when building artificial cognition, it is important to consider the full package of both the artificial intelligence inside the body interacting with the physical and social environment [4, 56].

In this chapter we provide an illustration of how an embodied perspective is used to imbue a robot with human-like skills. This requires a robot: we use the iCub platform, a child-sized humanoid robot specifically designed and built to facilitate developmental robotics [52] (see Fig. 5.1). In addition, an artificial cognitive model is required, which forms the theoretical backbone of the model.

Fig. 5.1 The iCub robot learning to map words to objects. In this experiment, iCub knows linguistic labels for three of the four objects. In response to the question where is the dax (dax is a novel word), it points at the unknown object. With this iCub demonstrates "fast-mapping", which is also observed in young children when they learn to map words to objects relying on only a few exposures and certain learning constraints [77]

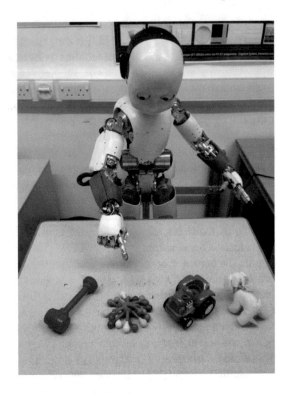

5.2.1   The Origins of Abstract Concepts and Number: A Detailed Study

Recent studies have proposed that multiple representational systems, involving both sensorimotor as well as linguistic systems, might be playing a primary role in how children acquire abstract concepts and words (e.g. [48]). Theories such as the LASS theory [11], according to which both the linguistic system as well as the sensorimotor system (through simulation) are activated in the processing of word meaning to different degrees under different task conditions, and the WAT (Words as Tools) approach proposed by Borghi and Cimatti [13], have suggested and furnished evidence on the synergetic role both language and sensorimotor experience play in the acquisition of abstract concepts, and on how important the modality by which these words are learned is.

Finger counting has been shown to be have an important role in the development of number cognition [3, 30]. Embodied cognition researchers find this innate ability particularly interesting, because of the sensorimotor contribution that it makes to the development of numerical cognition, and some consider it as "the most prominent example of embodied numerical cognition" [9]. Evidence coming from developmental, neurocognitive as well as neuroimaging studies suggest that finger counting activity helps build motor-based representations of number that continue to influence number processing well into adulthood, indicating that abstract cognition is rooted in bodily experiences [33]. These motor-based representations have been argued to facilitate the emergence of number concepts through a bottom-up process, starting from sensorimotor experiences [5].

In our view, finger counting, can also be seen as a means by which direct sensory experience can serve the purpose of grounding number words as well as numerical symbols, initially as low level symbols from the combination of already grounded ones, something known as grounding transfer [15, 40].

A number of connectionist models have simulated different aspects of number learning. A multi neural net approach was presented in [2] to explore quantification abilities and how they might arise in development, using a combination of supervised and unsupervised networks and learning techniques to simulate subitization (the phenomenon by which subjects appear to produce immediate quantification judgements, usually involving up to four objects, without the need to count them) and counting. The authors used a combined and modular approach, providing a simulation of different cognitive abilities that might be involved in the cognition of number (each of which would have their own evolutionary history in the brain), and is in keeping with Dehaenes triple code model [25]. In [60], using a hybrid artificial vision connectionist architecture, authors targeted aspects of language related to number such as linguistic quantifiers. They ground linguistic quantifiers such as few, several, many, in perception, taking into consideration contextual factors. Their model, after being trained and evaluated with experimental data using a dual-route neural network, is able to count objects in visual scenes and select the quantifier that best describes the scene.

Not many robotics studies have attempted to extend this. A cognitive robotics paradigm was used in [61, 62], where the authors explored embodied aspects of mathematical cognition such as the interactions between numbers and space, reproducing three psychological phenomena connected with number processing, namely size and distance effects, the SNARC effect and the Posner-SNARC effect.[1] The focus was on counting and on the contribution of counting gestures such as pointing. These models, however, did not consider the role of finger counting in numerical abilities.

Using a cognitive developmental robotics paradigm we explore whether finger counting and the association of number words (or tags) to the fingers, could serve to bootstrap the representation of number in a cognitive robot [23, 31, 32]. Our embodied robot experiments indicate that aspects of the development of this knowledge can be accounted for not only by way of bodily representations, but that a relatively simple artificial neural network is sufficient to achieve this.

The complete architecture proposed is shown in Fig. 5.2: the lower layer contains the motor controller/memory, and the auditory and the vision sub-systems. These are directly connected to the robotic platform. In the upper part there are the units

---

[1]SNARC, spatial-numerical association of response codes, is the effect whereby quantities seem to be spatially organised. People respond faster to small numbers with their left hand, and respond faster to large numbers with their right hand.

Fig. 5.2 Schematic of the robots cognitive system for number cognition

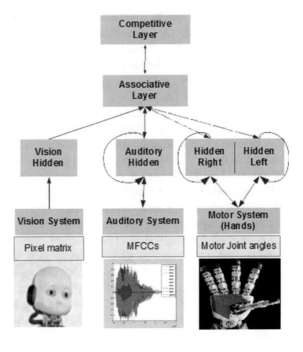

with abstract functions: the associative network and the competitive layer classifier. Note that the recurrent system's external inputs coincide with the outputs, indeed proprioceptive information from the motor and auditory systems is an input for the system during the training phase, while it is the control output when the system is operating.

Inputs are the joint angles, read from the encoders of the iCub hands, the mel-frequency cepstral coefficients (MFCC) to represent each number word from one to ten, and digits of 5×2 black and white pixels to represent number symbols. All numbers are in the range $[-1, 1]$. For number symbols, each element can be either -1, when the pixel is white, or $+1$ when the pixel is black.

The role of the competitive layer classifier is to simulate the final processing of the numbers, after a number is correctly classified into its class, the appropriate action can be started, e.g. the production of the corresponding word, of a symbol, the manipulation of an object and so on. The competitive layer classifier is implemented using the softmax transfer function that gives as output the probability/likelihood of each classification. We ensure that all of the output values are between 0 and 1, and that their sum is 1. The Switch/Associative Layer operates as a feedback system with the possibility to start and reset the motor/auditory layers and to derive the activations of one layer from the ones of the other.

Several experiments are run with the above architecture using the iCub robotic platform. In the first experiment [23], the main goal is to test the ability of the proposed cognitive system to learn numbers by comparing the performance of different ways of training the number knowledge of the robot with: (1) the internal representation (hidden units activation) of a given finger sequence; (2) the MFCC coefficients of number words out of sequence; (3) the internal representation of the number words sequence; (4) the internal representation of finger sequences plus the MFCC of number words out of sequence (i.e. learning words while counting); (5) internal representations of the sequences of both fingers and number words together (i.e. learning to count with fingers and words).

Looking at the developmental results, we again see that number words learnt out of sequence are the least efficient to learn. Conversely, if number words are learnt in sequence and internal representations are used as inputs, the learning is faster in terms of precision of classification, but is not as strong as when the learning involves also the use of fingers. Indeed, best results are obtained when internal representation of words and fingers are used together as input (Figs. 5.3 and 5.4).

A second experiment [31] focuses on learning associations between the internal representations (i.e. hidden unit activations) of number digits and the number words. Abstract concepts like the written representation of numbers is an important milestone in the childs unfolding cognitive development [81]. The young math learner must make the transition from a concrete number situation, in which the counting of objects (with fingers often being the first), to that of using a written form to stand for the quantities the sets of objects come to represent. This already challenging process

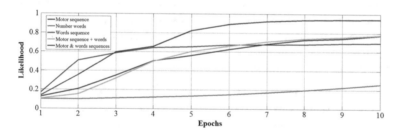

Fig. 5.3 Average likelihood with number classes with varying epochs

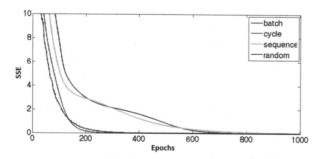

Fig. 5.4 Developmental learning of the association between number words and digits. Four different weight training methods are compared

is often coupled to that of learning a verbal number system, which depending on the particular language being used is not always transparent to children.

In this experiment four training strategies are considered: **Batch**, network weights are updated at the end of an entire pass through the input data; **Incremental** (3 strategies), network weights are learned with incremental updates after each presentation of an input order. Inputs are presented in **sequential** (i.e. from 1 to 10 each epoch), **random** (the order is randomly shuffled at each epoch), or **cyclic order** (the order is shifted after each epoch). Hidden unit activations are evaluated from the network with the best (lowest value of) performance function.

From this study we can conclude that the batch learning and the sequential strategies are less effective compared to others. They are slower to learn (i.e. they need more epochs) the final error (measured as sum of squared errors, or SSE) is several orders of magnitude higher than for random and cyclic order incremental training.

Once the number sequences are learnt, an interesting feature of the proposed cognitive system is the possibility to easily build up the ability to manipulate numbers with the development of the switch-associative network. Indeed, this ability can be modelled by extending the capabilities of the associative network from the simple start and stop, to its transferring and mapping to the basic operation of addition. The operation of addition can be seen as a direct development of the concurrent learning of the two recurrent units (motor and auditory). Indeed, if one of the two does the actual counting of the operands, the other can be used as a buffer memory to add the result, when it is done, the final number can be transferred from the buffer to the other unit and then inputted to the final processor (the classifier in our system). Here we want to build on this to show how the proposed architecture can take advantage of the previously learnt capability.

As an example let us consider 2+2, in this case the following steps will be taken:

1. The first operand is recognised by the visual system and, thanks to the associative network, the auditory internal representation is activated.
2. Auditory and motor networks will count until the corresponding activation of number 2 is reached. This step corresponds to the idle, start, counting (cycled twice) then done statuses of the associative network.
3. The sum operator is recognised so the associative layer resets the auditory network, while the first operand remains stored in the motor memory.
4. The second operand is recognised by the visual system, so the other networks restart counting as in step 1, until the auditory network reaches the activation corresponding to the number 2. In the meantime, the motor network reaches the activation of the number 4.
5. After the auditory network stops, the associative network recognises that the work is done so the total (4) is incepted from the fingers network to the auditory network thanks to the associative connection.
6. Finally the output of the resulting number (4) is produced for final processing (in our case the classifier).

The steps are depicted in Fig. 5.5.

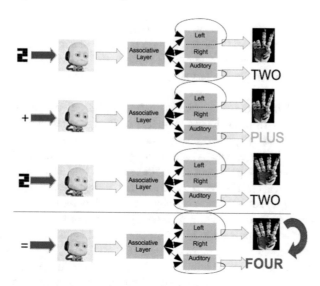

Fig. 5.5 Example of the execution of the operation of sum

5.3   Learning Through Social Interaction

As argued in the previous section, the seat of cognition is not the brain, but instead cognition emerges from the interaction between the brain, the body and the physical environment. While this holds for cognitive development of most animals, this picture is incomplete for some social species, and most significantly it is incomplete for humans. For human cognition to develop, one last element is required, namely the *social environment*. When including the social environment in cognition, this is sometimes known as "extended cognition"[2] [64].

While some elements of human cognition in all likelihood develop without input from the social environment (grasping and manipulation, for example, most likely develop without relying on social interaction), human infants grow up in a rich social environment. In this environment, social input in various shape and form is offered to the child, impacting on its cognitive development. Children learn from observing others: mimicry and imitation are potent mechanisms for acquiring cognitive skills [54] and rely on more skilled others to ostensibly demonstrate a skill, which is then imitated by the learner. Quite often the demonstration will be tailored as to promote successful interpretation of the demonstration by the young learner; demonstrating more slowly or emphasising salient elements of the skill to be acquired. The demon-strator also is able to provide feedback on the success of the demonstration, and can actively correct elements of the skill that are not yet fully established. Imitation, in

---

[2]Not to be confused with the Extended Mind hypothesis, in which cognition is argued to extend to the external world. As such external objects, such as canes, notepads and calculators, are seen as being integral to human cognition [21].

some form or other, is observed in many animals—primates and birds are known to imitate extensively—and as such imitation is a form of social learning that is not uniquely human [41]. However, *language is uniquely human*. While many species communicate, no other species has access to the open communication system that language is.

It has been claimed that language is such a hard problem that is unlearnable, and can only result from an innate language of thought pre-specified by genetic evolution [19, 35, 36, 58]. How else could the child, a passive observer viewing a cluttered scene, know to which feature or collection of features a spoken word referred? By contrast, embodiment views the learning child as anything but passive [73]; their attention is clearly focused and they are 'doing' (reaching, holding, banging, manipulating…) sometimes being physically lead by the caregiver [82]. Words are not merely spoken either, child directed speech is not simply speech directed at the child, but is manipulated to highlight events, and is just as much directed by the child's attention and reaction. Smith et al. [68] go further, highlighting just how dominant a held object is in the infants' field of view. From this perspective the language learning child is not really aware of all the perceptual clutter (the held object is simply occluding most of it), and the spoken words often relate to what the child is currently doing, holding or attending to. As such, the learning of simple concrete item-based word-object and word-action mappings becomes possible, and we have demonstrated the basic principles involved on robotic systems [53, 77].

Moving beyond simple word-object mappings, Tomasello [75] further highlights that from a concrete item-based vocabulary children gradually (over many years) develop the ability to construct more abstract and adult-like linguistic constructions. This gradually increasing complexity of language presents a significant challenge to the hypothesis that language is innate. We therefore suggest that language is not symbol manipulation in the head but is a sensorimotor process, whereby words prime or predict associated features (be they combinations of sensory features, motor actions, or affordances), and likewise these features prime their associated words.

Language has many functions, next to its obvious function as communication system, it also supports cognition in ways that are not always recognised. One is that language is used in concept acquisition. Humans are, in the words of Terrence Deacon, a "symbolic species" [24]. We cut up our perceptual experience in concepts, and can subsequently order these concepts into hierarchical concepts. Concepts allow us to reason and are the brain's way of compressing sensory input into fewer, finite units. When we assign a linguistic label to a concept (a word, utterance or linguistic construction), that concept can be communicated to others. Concepts are central to human cognition, but it is not clear where concepts come from. Are they learnt by a child when growing up? Are some concepts innate, and some learnt? If so, which concepts are innate and which are learnt? And, when concepts are acquired, how exactly are they acquired?

This latter question is important: *how can a child acquire concepts*? And by extension, can similar processes be used to let an artificial system, such as a robot, acquire human-like concepts? Sect. 5.2.1 shows how linguistic labels (i.e. words) can be used to bind external perception with internal representations. In this section we look into the contribution of the social interaction on word-meaning acquisition.

As children develop, there is a rich and frequent interaction between the child and the environment. Not only the physical environment is explored, but also the social environment. And while the physical environment does not actively respond to actions by the child, the social environment (i.e. the child's siblings, parents or other carers) do actively respond to the developing child. In language learning, phenomena such as infant-directed speech—where carers address the child in a simpler and hyperarticulated language—aids language acquisition. Likewise, when learning the semantics of language, the carer-child dyad often engages in rich interactions in which joint attention and deictic pointing is combined with the naming of objects or actions. Together with a number of learning biases [76], this enables the child to rapidly acquire a set of words and semantic associations [50].

Inspired by these observations, we explore if similar mechanisms could be used to accelerate robot learning. In our study, the robot learns from people in a way that is similar to how children learn from others. In this *socially guided machine learning* [47] the machine is not offered a batch of training data to learn from, but instead engages in a high-resolution interaction which a human, whereby the machine invites the human to offer tailored training data to optimise its learning.

We focus on the task of learning associations between words and referents [12], for which the learner has to construct internal representations linking linguistic symbols with external referents. These dynamics of meaning acquisition have been explored in detail, often using simulations in which agents bootstrap a shared symbol system and meanings—e.g. [70–72]. However, in these simulations the agents do not actively influence the learning process by querying their social environment. Early simulations have shown that active learning can result in improved performance [29]. When an agent uses strategies to actively elicit better training examples from other agents, the learner learns faster and better. The strategies consist of active learning (whereby the learner points out a referent in the world which it would like to know the linguistic label for, similar to a child pointing out something in the presence of a carer), knowledge querying (whereby the learner verifies its internal knowledge by using it and asking the carer for feedback, mimicking the way in which children name objects in their environment and invite adults to correct them or confirm their linguistic labels) and contrastive learning (in which an association between a word and a referent is increased, but association between that word and other referents is decreased).

While the strategies result in better learning in simulation, we wish to confirm if this would still hold in the real world: where a social robot is learning from a human tutor. To this end we design a setup in which a social robot sits across a human tutor (see Fig. 5.6). Between the robot and human, a touchscreen is placed through which the interaction takes place. The participants are asked to teach the robot the concepts of animal classes (mammal, insect, invertebrate, …), using images of animals (e.g. a bear, an ant, a lizard, …).

The robot uses learning strategies identical to the simulation model [29], the contribution of the social robot setup is on the one hand the learning from people

Fig. 5.6 Experimental setup, with the social robot sitting across the participant. The participant is invited to teach the robot to correctly match images of animals with animal classes

rather than from other simulated agents, and on the other hand the introduction of additional communication channels, such as eye gaze and affective communication. To aid social communication and to invite people to help the robot learn, the robot is deliberately designed to resemble a young child [26].

The experiment uses two conditions: in one condition people interact with a social robot, using the learning strategies detailed above and using congruent linguistic and facial expressions to support the active learning, in the second condition, the robot learns, but does not use any of the above strategies to learn more efficiently; we refer to this condition as the "non-social robot". 19 subjects interacted with the social robot, and 20 with the non-social robot condition. Full details can be found in [28].

Results show that in both conditions, the robot learns to correctly match instances of animals to animal classes, illustrating that the learning algorithms works as expected. The social robot learning is faster and slightly better than that of the non-social robot, as predicted by the simulation results. It is interesting to observe that there is a marked gender effect in the results: female participants achieve a significantly higher learning success when interacting with the social robot, and this drops significantly for the non-social robot. Male participants achieve similar learning results for both conditions (see Fig. 5.7). This suggests that female participants in our study are more sensitive to the social cues expressed by the robot, while this is not the case for the male participants.

Finally, a careful analysis of the data shows that people readily form a "mental model" of the robot: both in the social and non-social condition people will offer training data to the robot that are tailored to the current performance of the robot [28], thereby showing that people form a model of the robot's mental state.

Fig. 5.7 Word-meaning learning of the social and non-social robot; note how the social robot learns more from female participants in our experiment, while it learns significantly less from male participants. Error bars are 95 % confidence intervals

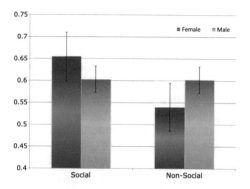

This experiment convincingly illustrates that social robots can elicit better training experiences. The careful design of the appearance and the behaviour of the robot can lead to improved learning on robots, and taps into the human propensity for tutoring.

5.4 Powering Artificial Cognition with Spiking Neural Networks

The desire to endow robots with sophisticated human-like capabilities raises some major challenges as traditional computing and engineering approaches can only achieve so much. They can and have been used to mimic human capabilities to various levels of abstraction but it is difficult to make artificial systems that behave in the same way as natural ones do if we do not fully understand all the neural processing which generates our own behaviour. It is also often difficult to translate biological concepts into a traditional computing/engineering framework without making severe compromises. The sensory pre-processing and higher level cognitive processing that is required to achieve such human-like learning capabilities in an embodied developmental robotics scenario likely requires significant computing power which is in conflict with the limited energy resources available on an autonomous robot. It should be noted, however, that natural neural systems manage to operate in real time, be fault tolerant and flexible despite having very low power requirements. Therefore, it seems logical to explore more in depth bio-inspired approaches to robotics. For example, where artificial brains and nervous systems are implemented using techniques inspired by greater understanding of how real neurons work. Arbib et al. [6] defines the field of Neurorobotics as

> … the design of computational structures for robots inspired by the study of the nervous system of humans and other animals.

and suggests that neural models more closely matching the biology may more clearly reveal the computational principles necessary for cognitive robotics while illuminating human (and animal) brain function.

In parallel, the field of Computational Neuroscience (the study of brain function using biologically realistic models of neurons over multiple scales from single neuron dynamics up to networks of neurons) has made considerable progress on spiking neuron based models of sensory and cognitive processes in the mammalian neo-cortex. Spiking Neural Networks (SNNs) are the "third generation" of Neural Networks [49] and mimic how real neurons compute: with discrete pulses rather than a continuously varying activation. The spiking neuron is, of course, still an abstraction from a real neuron but depending upon the application and required level of biological detail, there are various types of spiking neuron model to choose from. However, there is also a trade-off between the level of biological detail and computational overhead (for a review and discussion see [42]).

In neurobiological experimental studies neuron responses have been predominantly measured as a spike rate, however there is accumulating evidence that spike timing is also important. Experimental evidence exists for fast processing (occurring within 100 ms of an image presentation) in the human visual system [74] which implies that spike timing information may be more important than spike rates as there is not enough time to generate a meaningful spike rate in very short time intervals. Spike timing also seems to be important in learning: Spike Timing Dependent Plasticity (STDP) is a currently favoured model for learning in real neurons. Experimental and modelling studies have shown that this form of Hebbian plasticity, where the relative firing times of pre and post-synaptic neurons influence the strengthening or weakening of connections, is the mechanism that real neurons use [69]. When firing times are causally related (i.e. the pre-synaptic spike is emitted before the post-synaptic spike) then the synapse is strengthened (Long Term Potentiation or LTP). When firing times are not causally related (i.e. the post-synaptic spike occurs before the pre-synaptic one) then the synapse is weakened (Long Term Depression or LTD).

Of particular relevance to modelling human cognitive function, some neurobiological experiments have suggested that spike-timing is also directly important at the cognitive/behavioural level as well as in learning [8, 66].

There have been a few research projects involving robotic implementations based upon human-like capabilities using spiking neural networks. Three notable examples are the Darwin series of robots [34, 44], the humanoid CRONOS/SIMNOS project [38] and the control of an iCub arm with an SNN and STDP [14].

The iSpike API [39] has a lot of promise for facilitating interfacing between SNNs and humanoid robots but as yet no practical demonstrations exist. Therefore, it is only relatively recently that works using spiking neural networks for practical humanoid robotics applications have begun to emerge. Certainly advances in software and hardware over the last ten years or so have made SNNs an increasingly feasible option for robotics applications. On the software side several general purpose spiking neuron simulators are freely available which means that researchers do not have to code a modelling framework from scratch, and they also benefit from a community of users using the same tool. Desktop computing hardware is now available that can perform parallel processing (e.g. GPU) at an affordable price. But this can only take us so far. Until now, in practice most Neurorobotic systems, e.g. Chersi [18] have simulated the neural component on an external host PC which limits the ability of the robot to truly perform autonomously in real time.

The emerging field of Neuromorphic Engineering is making it possible to simulate large neural networks in hardware in real time. Neural chips are massively parallel arrays of processors that can simulate thousands of neurons simultaneously in a fast, energy efficient way, thus making it possible to move neural applications on board robots. This technology is currently being employed in dedicated hardware devices to perform specific bio-inspired functions, for example, the asynchronous temporal contrast silicon retina [27] and the silicon cochlea [17]. There have also been several larger-scale projects for general purpose brain modelling. For example, the CAVIAR project; a massively parallel hardware implementation of a spike-based sensing-processing-learning-actuating system inspired by the physiology of the nervous system [65], the FACETS project (completed in 2010) delivering both neuromorphic hardware and software and the NeuroGrid project at Stanford which has developed a hybrid analogue-digital neuromorphic solution capable of modelling up to 1 million neurons (reviewed in [67]). More recently, the SpiNNaker project has delivered a state-of-the-art real-time neuromorphic modelling environment that can be scaled-up to model up to a billion point-neuron models [43].

The parallel advances in computational neuroscience and in the hardware implementation of large-scale neural networks, provide the opportunity for an accelerated understanding of brain functions and for the design of interactive robotic systems based on brain-inspired control systems. However, currently there are very few practical robotics implementations using neuromorphic systems. Two notable works are [46] which developed a solution using both a silicon retina, an FPGA and neuromorphic hardware to enable a humanoid robot to point in the direction of a moving object, and, more recently [22] which developed a line following robot using a silicon retina and a prototype 4-chip SpiNNaker neuromorphic board.

Adams et al. [1] recently introduced a Neurorobotics system integrating the humanoid iCub robot and a SpiNNaker neuromorphic board to solve a behaviourally relevant task: goal-directed attentional selection. Using an enhanced version of an existing SNN model with layers inspired by real brain areas in the mammalian visual system [37], iCub was equipped with the ability to fixate attention upon a selected stimulus. Although in this particular implementation the selected or "preferred" stimulus was fixed in advance the network has the option to enable STDP learning to learn the preferred stimulus.

This study demonstrated the first steps in creating a cognitive system incorporating several important features for prospective Neurorobots:

1. Universally configurable hardware that can run a variety of SNNs.
2. Standard interfacing methods that eliminate difficult low-level issues of connectors, cabling, signal voltages, and protocols.
3. Scalability—the SpiNNaker hardware is designed to be able to run very large SNNs and the optimal placement of networks onto the hardware is abstracted away from the user.

More work needs to be done to develop practical applications that have a solid biologically-inspired theoretical basis and which can be scaled up and transferred seamlessly to run on neuromorphic hardware to take advantage of their special-ist processing capabilities and low power requirements. For realistically large and effective SNNs to become possible in robotic hardware, ensuring that future neural models and simulations are actually implementable in neuromorphic hardware is important. It is also important to develop models which challenge the capabilities of such hardware and stimulate further developments.

5.5 Conclusion and Outlook

The studies described here illustrate how artificial cognition, just as its natural coun-terpart, benefits from being grounded and embodied. This occurs at several levels: the body of the cognitive system shapes its cognition, but so does the physical environ-ment and the social environment. Human-like cognition results from the tight interac-tion between these four constituents, see Fig. 5.8. When one of the four constituents is missing, it is still possible to recreate certain aspects of cognition. For example, the social component is missing in much of animal cognition and some aspects of human cognition—such as manipulation or locomotion—can develop in the absence of the social environment. Or when the body is missing, systems have been shown to still be able to reach human levels of performance on specific tasks. Latent Semantic Analy-sis, for example, is able to pass synonymy tests just using statistical co-occurrence information of words in large text corpora [45]. However, to replicate natural human cognition in its full scope, we make argue that all four components—body, brain,

Fig. 5.8 The four constituents of human cognition: the brain, the body, and the physical and social environment. We argue that no human-like cognition can develop without the presence of these four components

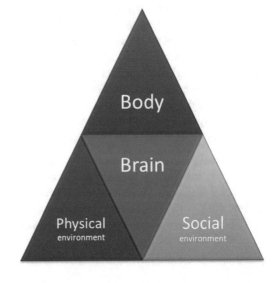

physical environment and social environment—are required, and that all four cannot be seen as separate entities, rather they intertwine and operate in close association with each other. In addition, we believe that a thorough understanding of the neural processes underpinning natural cognition will aid in the design and implementation of artificial equivalents; key to this might be spiking neural networks.

References

1. Adams S, Rast A, Patterson C, Galluppi F, Brohan K, Perez-Carrasco JA, Wennekers T, Furber S, Cangelosi A (2014) Towards real-world neurorobotics: Integrated neuromorphic visual attention. In: Proceedings of 21st international conference on neural information processing (ICONIP), pp 563–570
2. Ahmad K, Casey M, Bale T (2002) Connectionist simulation of quantification skills. Connect Sci 14(3):165–201
3. Alibali MW, DiRusso AA (1999) The function of gesture in learning to count: more than keeping track. Cognit Dev 14(1):37–56
4. Anderson ML (2003) Embodied cognition: a field guide. Artif Intell 149(1):91–130
5. Andres M, Di Luca S, Pesenti M (2008) Finger counting: the missing tool? Behav Brain Sci 31(06):642–643
6. Arbib MA, Metta G, van der Smagt P (2008) Neurorobotics: From vision to action. In: Khatib O Siciliano B (eds) Springer Handbook of Robotics, Springer-Verlag, pp 1453–1480
7. Asada M, Hosoda K, Kuniyoshi Y, Ishiguro H, Inui T, Ogino Y, Yoshida C (2009) Cognitive developmental robotics: a survey. IEEE Trans Auton Mental Dev 1(1):12–34
8. Ayzenshtat I, Meirovithz E, Edelman H, Werner-Reiss U, Bienenstock E, Abeles M, Slovin H (2010) Precise spatiotemporal patterns among visual cortical areas and their relation to visual stimulus processing. J Neurosci 40:11232–11245
9. Bahnmueller J, Dresler T, Ehlis AC, Cress U, Nuerk HC (2014) Nirs in motionunraveling the neurocognitive underpinnings of embodied numerical cognition. Front Psychol 5:743
10. Barsalou LW (2008) Grounded cognition. Annu Rev Psychol 59:617–645
11. Barsalou LW, Santos A, Simmons WK, Wilson CD (2008) Language and simulation in conceptual processing. Symbols, embodiment, and meaning pp 245–283
12. Bloom P (2000) How children learn the meanings of words. The MIT Press, Cambridge
13. Borghi AM, Cimatti F (2012) Words are not just words: the social acquisition of abstract words. Rivista Italiana di Filosofia del Linguaggio 5:22–37
14. Bouganis A, Shanahan M (2010) Training a spiking neural network to control a 4-dof robotic arm based on spike timing-dependent plasticity. Proc IJCNN 2010:1–8
15. Cangelosi A, Riga T (2006) An embodied model for sensorimotor grounding and grounding transfer: experiments with epigenetic robots. Cognit Sci 30(4):673–689
16. Cangelosi A, Schlesinger M (2015) Developmental robotics: from babies to robots. The MIT Press, Cambridge
17. Chan V, Liu SC, van Shaik A (2007) Aer ear: a matched silicon cochlea pair with address event representation interface. IEEE Trans Circuits Syst I: Spec Issue Smart Sens 54:48–49
18. Chersi F (2012) Learning through imitation: a biological approach to robotics. IEEE Trans Auton Mental Dev 4(3):204–214
19. Chomsky N (1995) The minimalist program. Cambridge Univ Press, Cambridge
20. Christaller T (1999) Cognitive robotics: a new approach to artificial intelligence. Artif Life Robot 3(4):221–224
21. Clark A, Chalmers D (1998) The extended mind. analysis pp 7–19
22. Davies S, Patterson C, Galuppi F, Rast A, Lester D, Furber S (2010) Interfacing real-time spiking i/o with the spinnaker neuromimetic architecture. In: Proceedings 17th international

conference, ICONIP 2010: 17th international conference, ICONIP 2010: Australian journal of intelligent information processing systems, vol 11, pp 7–11

23. De La Cruz VM, Di Nuovo A, Di Nuovo S, Cangelosi A (2014) Making fingers and words count in a cognitive robot. Front Behav Neurosci 8:1–12

24. Deacon TW (1997) The symbolic species: the co-evolution of language and the brain. Norton, New York

25. Dehaene S (2000) The cognitive neuroscience of numeracy: exploring the cerebral substrate, the development, and the pathologies of number sense. Scientific research faces a new millennium, Carving our destiny

26. Delaunay F, de Greeff J, Belpaeme T (2010) A study of a retro-projected robotic face and its effectiveness for gaze reading by humans. Proceedings of the 5th ACM/IEEE international conference on human-robot interaction (HRI2010), Mar 2–5 (2010). IEEE Press, Osaka, Japan, pp 39–44

27. Delbruck T (2008) Frame-free dynamic digital vision. In: Proceedings of international advanced electronics for quality life and society, symposium on secure-life electronics, pp 21–26

28. de Greeff J, Belpaeme (2015) Why robots should be social: Enhancing machine learning through social human-robot interaction. PLOS One In press

29. de Greeff J, Delaunay F, Belpaeme T (2009) Human-robot interaction in concept acquisition: a computational model. In: Triesch J, Zhang Z (eds) IEEE international conference on development and learning (ICDL 2009). IEEE, Shanghai

30. Di Luca S, Pesenti M (2011) Finger numeral representations: more than just another symbolic code. Front Psychol 2:272

31. Di Nuovo A, De La Cruz VM, Cangelosi A (2014a) Grounding fingers, words and numbers in a cognitive developmental robot. In: IEEE symposium on computational intelligence, cognitive algorithms, mind, and brain (CCMB, 2014). IEEE, pp 9–15

32. Di Nuovo A, De La Cruz VM, Cangelosi A, Di Nuovo S (2014b) The icub learns numbers: An embodied cognition study. In: International joint conference on neural networks (IJCNN, 2014). IEEE, pp 692–699

33. Domahs F, Moeller K, Huber S, Willmes K, Nuerk HC (2010) Embodied numerosity: implicit hand-based representations influence symbolic number processing across cultures. Cognition 116(2):251–266

34. Edelman G (2007) Learning in and from brain-based devices. Science 318:1103–1105

35. Fodor JA (1975) The language of thought. Harvard University Press, Cambridge

36. Fodor JA (2008) LOT 2: the language of thought revisited: the language of thought revisited. Oxford University Press, Oxford

37. Galluppi F, Brohan K, Davidson S, Serrano-Gottarredona T, Corasco JAP, Linares-Barranco B, Furber S (2012) A real-time, event driven neuromorphic system for goal-directed attentional selection. In: ICONIP 2012

38. Gamez D, Newcombe R, Holland O, Knight R (2006) Two simulation tools for biologically inspired virtual robotics. In: Proceedings of the IEEE 5th chapter conference on advances in cybernetic systems

39. Gamez D, Fidjeland A, Lazdins E (2012) Iispike: a spiking neural interface for the icub robot. Bioinspir Biomim 7(2):025008

40. Harnad S (1990) The symbol grounding problem. Phys D: Nonlinear Phenom 42(1):335–346

41. Hurley SL, Chater N (2005) Perspectives on Imitation: mechanisms of imitation and imitation in animals, vol 1. MIT Press

42. Izhikevich EM (2004) Which model to use for cortical spiking neurons? IEEE Trans Neural Netw 15(5):1063–1070

43. Jin X, Lujan M, Plana L, Davies S, Temple S, Furber S (2010) Modeling spiking neural networks on spinnaker. Comput Sci Eng 21(5):91–97

44. Krichmar J, Edelman G (2003) Brain-based devices: intelligent systems based on principles of the nervous system. In: Proceedings of the 2003 IEEE/RSJ international conference on intelligent robots and systems vol 1

45. Landauer TK, Dumais ST (1997) A solution to plato's problem: the latent semantic analysis theory of acquisition, induction, and representation of knowledge. Psychol Rev 104(2):211–240
46. Linares-Barranco A, Gomez-Rodriguez F, Jimenez-Fernandez A, Delbr-ck T, Lichtensteiner P (2007) Using fpga for visuo-motor control with a silicon retina and a humanoid robot. In: Proceedings of the IEEE symposium on circuits and cystems (ISCAS 2007), pp 1192–1195
47. Lockerd A, Breazeal C (2004) Tutelage and socially guided robot learning. In: Proceedings of IEEE/RSJ international conference on intelligent robots and systems (IROS 2004)
48. Louwerse MM, Jeuniaux P (2010) The linguistic and embodied nature of conceptual processing. Cognition 114(1):96–104
49. Maass W (1997) Networks of spiking neurons: the third generation of neural network models. Neural Netw 10:1659–1671
50. Markman EM (1989) Categorization and naming in children: problems of induction. The MIT Press, Cambridge
51. Meltzoff AN, Moore MK (1994) Imitation, memory, and the representation of persons. Infant Behav Dev 17(1):83–99
52. Metta G, Sandini G, Vernon D, Natale L, Nori F (2008) The icub humanoid robot: an open platform for research in embodied cognition. In: Proceedings of the 8th workshop on performance metrics for intelligent systems, ACM, pp 50–56
53. Morse AF, Belpaeme T, Cangelosi A, Floccia C, Carlson L, Hoelscher C, Shipley T (2011) Modeling u-shaped performance curves in ongoing development. In: Expanding the space of cognitive science: proceedings of the 23rd annual meeting of the cognitive science society
54. Nehaniv CL, Dautenhahn K (2002) Imitation in animals and artifacts. MIT Press, Cambridge
55. Pecher D, Zwaan RA (2005) Grounding cognition: The role of perception and action in memory, language, and thinking. Cambridge University Press, Cambridge
56. Pfeifer R, Bongard J (2006) How the body shapes the way we think: a new view of intelligence. MIT press, Cambridge
57. Pfeifer R, Scheier C (1999) Understanding intelligence. The MIT Press, Cambdrige
58. Pinker S (1994) The language instinct: how the mind creates language. W. Morrow, New York
59. Pulvermüller F (2005) Brain mechanisms linking language and action. Nat Rev Neurosci 6(7):576–582
60. Rajapakse RK, Cangelosi A, Coventry KR, Newstead S, Bacon A (2005) Connectionist modeling of linguistic quantifiers. In: Artificial neural networks: formal models and their applications-ICANN 2005, Springer, pp 679–684
61. Rucinski M, Cangelosi A, Belpaeme T (2011) An embodied developmental robotic model of interactions between numbers and space. Expanding the space of cognitive science: proceedings of the 23rd annual meeting of the cognitive science society. Cognitive Science Society Austin, TX, pp 237–242
62. Rucinski M, Cangelosi A, Belpaeme T (2012) Robotic model of the contribution of gesture to learning to count. In: IEEE International conference on development and learning and epigenetic robotics (ICDL, 2012). IEEE, pp 1–6
63. Samuelson LK, Smith LB, Perry LK, Spencer JP (2011) Grounding word learning in space. PLOS One 6(12):e28,095
64. Seabra Lopes L, Belpaeme T, Cowley S (2008) Beyond the individual: new insights on language, cognition and robots. Connect Sci 20(4):231–237
65. Serrano-Gotarredona R, Oster M, Lichtsteiner P, Linares-Barranco A, Paz-Vicente R, Gomez-Rodriguez F, Camunas-Mesa L, Berner R, Rivas M, Delbr-ck T, Liu SC, Douglas R, Hafliger P, Jimenez-Moreno G, Civit A, Serrano-Gotarredona T, Acosta-Jimenez A, Linares-Barranco B (2009) Caviar: A 45k-neuron, 5m-synapse, 12g-connects/sec aer hardware sensory-processing-learning-actuating system for high speed visual object recognition and tracking. IEEE Trans Neural Netw 20:1417–1438
66. Shmiel T, Drori R, Shmiel O, Ben-Shaul Y, Nadasdy Z, Shemesh M, Teicher M, Abeles M (2006) Temporally precise cortical firing patterns are associated with distinct action segments. J Neurophysiol 96:2645–2652

67. Silver R, Boahen K, Grillner S, Kopell N, Olsen K (2007) Neurotech for neuroscience: unifying concepts, organizing principles, and emerging tools. J Neurosci 27:11,807–11,819
68. Smith LB, Yu C, Pereira AF (2011) Not your mothers view: the dynamics of toddler visual experience. Dev Sci 14(1):9–17
69. Song S, Miller KD, Abbott LF (2000) Competitive hebbian learning through spike-timing dependent synaptic plasticity. Nat Neurosci 3:919–926
70. Steels L (2003) Evolving grounded communication for robots. Trends Cognit Sci 7(7):308–312
71. Steels L, Belpaeme T (2005) Coordinating perceptually grounded categories through language. A case study for colour. Behav Brain Sci 24(8):469–529
72. Steels L, Kaplan F, McIntyre A, Van Looveren J (2002) Crucial factors in the origins of word-meaning. In: Wray A (ed) The transition to language. Oxford University Press, Oxford, pp 252–271
73. Thelen E, Smith LB (1998) Dynamic systems theories. Handbook of child psychology
74. Thorpe S, Fize D, Marlot C (1996) Speed of processing in the human visual system. Nature 381:520–522
75. Tomasello M (2000) The item-based nature of childrens early syntactic development. Trends Cognit Sci 4(4):156–163
76. Tonkes B, Willes J (2002) Minimally biased learners and the emergence of language. In: Wray A (ed) The transition to language. Oxford University Press, Oxford
77. Twomey K, Morse A, Cangelosi A, Horst J (2014) Competition affects word learning in a developmental robotic system. In: 14th neural computation and psychology workshop
78. Vernon D (2014) Artificial cognitive systems: a primer. MIT Press, Cambridge
79. Vernon D, Metta G, Sandini G (2007) A survey of artificial cognitive systems: implications for the autonomous development of mental capabilities in computational agents. IEEE Trans Evolut Comput 11(2):151–180
80. Warneken F, Chen F, Tomasello M (2006) Cooperative activities in young children and chimpanzees. Child Dev 77(3):640–663
81. Zhou X, Wang B (2004) Preschool childrens representation and understanding of written number symbols. Early Child Dev Care 174(3):253–266
82. Zukow-Goldring P, Arbib MA (2007) Affordances, effectivities, and assisted imitation: caregivers and the directing of attention. Neurocomputing 70(13):2181–2193

Chapter 6
Going Further in Affective Computing: How Emotion Recognition Can Improve Adaptive User Interaction

Sascha Meudt, Miriam Schmidt-Wack, Frank Honold, Felix Schüssel, Michael Weber, Friedhelm Schwenker and Günther Palm

Abstract This article joins the fields of emotion recognition and human computer interaction. While much work has been done on recognizing emotions, they are hardly used to improve a user's interaction with a system. Although the fields of affective computing and especially serious games already make use of detected emotions, they tend to provide application and user specific adaptions only on the task level. We present an approach of utilizing recognized emotions to improve the interaction itself, independent of the underlying application at hand. Examining the state of the art in emotion recognition research and based on the architecture of *Companion*-System, a generic approach for determining the main cause of an emotion within the history of interactions is presented, allowing a specific reaction and adaption. Using such an approach could lead to systems that use emotions to improve not only the outcome of a task but the interaction itself in order to be truly individual and empathic.

Sascha Meudt, Miriam Schmidt-Wack, Frank Honold and Felix Schüssel—These authors contributed equally to this contribution.

S. Meudt · M. Schmidt-Wack (✉) · F. Schwenker · G. Palm
Institute of Neural Information Processing, Ulm University, Ulm, Germany
e-mail: sascha.meudt@uni-ulm.de

M. Schmidt-Wack
e-mail: miriam.k.schmidt@uni-ulm.de

F. Schwenker
e-mail: friedhelm.schwenker@uni-ulm.de

G. Palm
e-mail: guenther.palm@uni-ulm.de

F. Honold · F. Schüssel · M. Weber
Institute of Media Informatics, Ulm University, Ulm, Germany
e-mail: frank.honold@uni-ulm.de

F. Schüssel
e-mail: felix.schuessel@uni-ulm.de

M. Weber
e-mail: michael.weber@uni-ulm.de

© Springer International Publishing Switzerland 2016
A. Esposito and L.C. Jain (eds.), *Toward Robotic Socially Believable Behaving Systems - Volume I*, Intelligent Systems Reference Library 105, DOI 10.1007/978-3-319-31056-5_6

6.1 Introduction

In recent years, interaction with computer systems has become ubiquitous as already predicted by Mark Weiser in 1991 [79]. Today, people are surrounded by computers throughout their daily lives, be it in the form of smartphones, different components in modern cars, smart TVs, and even home appliances like vacuum robots or internet connected fridges. Information technologies and services are everywhere, not only in everyday life activities, but in the workplace as well. Human-computer interaction (HCI) as a scientific discipline is not restricted to systems used by few experts, but has influence on almost every human being, at least in modern civilizations. Therefore, HCI has the responsibility to create usable, universal and useful technologies as Ben Shneiderman states in his foreword in [71]. This can be achieved by building systems that "... empower individuals, enrich human relationships, encourage cooperation, build trust, and support empathy" (ibid.).

In this article, we focus on the closely related aspects of *individuality* and *empathy* in HCI. A system can be attributed to be user-individual, if it takes into account the user's particular abilities, preferences, requirements as well as the user's current needs. In addition, an empathic system can be considered as being aware of a user's situation and emotional (affective) state, and is able to adapt via diverse established system components (cf. [5]), even on different levels. This kind of systems are also called *Companion*-Systems [80] and consists of different components which are able to continuously adapt to the context of use (CoU) [30, 36]. This context [21] comprises (amongst others things) a user model and an environment model [35].

In order to adapt to and deal with the user, the system has to record sensory inputs via cameras, microphones or even measuring tools for bio-physiological data and derive the user's explicit messages and equally important the implicit cues. The ability to recognize this affective state of the communication partner is not only essential in human-human interaction (HHI) but also in HCI [17, 55].

However, there are severe problems concerning the solution of such a multi-modal recognition task. Most of the available datasets for training consist of acted emotions [14, 76], which are often more expressive than emotions in real-life scenarios. If users are recorded in a more natural environment, emotional expressions occur quite rarely [20]. In addition, there exist different psychological emotion models such as *Basic Emotions* [23] or *Valence-Arousal-Dominance* [61], which can be used as background for the labeling process. But it is still an unanswered question, which model fits best or if it is more promising to use descriptions, as for example *(un-)challenged* or *(un-)motivated*.

The indicators for the user's affective state can have their origin in the video channel: facial expression [13, 24], body posture [4, 27] as well as hand gestures [12, 51] provide adequate information. In addition, audio signals can deliver information concerning the user's emotional state [54, 59], and even detection of laughter can give some insights in the user's condition [11]. The third source for emotional indication is the bio-physiological channel. Features extracted from heart rate or skin conductance can be used to estimate the user's affective state by utilizing appropriate tools for measurements [56]. These manifold features are then utilized as input for a classification model to recognize the occurring emotions.

The use of emotions in HCI is manifold. In terms of affective computing [55] and with focus on the systems' possible output, systems are able to express diverse emotional states in order to simulate empathy. The thereby applied modalities range from simple emoticons to complex live-like virtual representations using multiple modalities [3] or solid embodiments [6]. Emotional expressions can also be used to support non-verbal communication [25, 33]. Additionally, emotions can be interpreted as implicit user inputs for further reasoning of adaptive system behavior.

According to [9] the effects of emotional affect have influence on a person's attention, memory, performance, and assessment of situations. Therefore, these effects are addressed in serious games or e-learning software [60]. The user can be supported in his educational progress via an individual adaptation of the system's applied strategy of education, or even by adapting the course or content of the dialog (e.g. by praising, criticizing, or encouraging). Thereby, the decision for adaptation can be attributed to the user's current task performance (e.g. time to completion) but also to a recognized emotion, as described in [52].

In the last decades, humans gained insights into the complex topic of emotion, and we learned to distinguish emotion from related terms like mood, and sentiment. Emotions have been identified to be object-directed, as they "imply and involve relationships with a particular object" [26]. That's why emotions can be seen as a reaction to a specific situation and their strongest effect is to capture attention [9].

As stated by [44], not only emotions but also affective states are of interest in HCI. According to them, it is important to know if a user is "*satisfied, more confused, frustrated*, or simply *sleepy*", in order to provide additional feedback or to proceed with the current way of interaction. The authors of [44] go on with a description of possible influences of emotions on different concepts as e.g. focus and attention, perception, memorizing and learning, motivation and performance, goal-generation, decision making, strategic planning, and communication. All these concepts are important to address, if a system shall be able to act as *Companion*-System.

We think of *Companion*-Systems as smart systems, which are able to recognize emotions on the one hand, and to react on them by adaptation on different levels on the other hand (see Fig. 6.7). This follows a still ongoing challenge, as motivated in [44] with the principle "that computers should be adapting to people rather than vice versa." But these possibilities for adaptation on different levels come with the decision problem, which is concerned with the reasoning about the right level of adaptation, and thus with the emotion's related object. If, for example, a user dislikes the current user interface (e.g. the use of a smartphone screen in bright sunshine), it is unnecessary to adapt on any other levels. The suggestion of a different task, or the adaptation of the currently used dialog structure won't be as effective, as e.g. offering an additional speech interface in such a situation.

In the remainder of this article, we focus on *Companion*-Systems with their ability to perceive affect, but without the ability to express affect in terms of an avatar. This does not mean that affect does not occur in the system's output at all, but it is rather subtle (e.g. with the use of an adaptive dialog structure using terms for excuses (cf. [1]) and encouragements, or with the use of emotion-adaptive background music). According to Picard's "four categories of affective computing" [55], we locate the named *Companion*-System in the fourth category, as a systems which is able to perceive *and* express affect.

To the best of our knowledge there is no system or approach, that meets the following three contributions in combination: (1) enabling a system to sense emotions using multi-modal cues, (2) offer the ability to infer the emotion-related cause of a sensed change in the user's affective potential, and (3) to react in an adequate way on the cause-related architectural layer with an adaptation of the used strategy for input and/or output.

6.2   How to Classify Emotional States

As humans have to learn to interpret the emotional states of their fellows from childhood on, a *Companion*-System needs to gain this ability, too. The methods of machine learning can give this kind of advantage to computers. This approach is data driven and needs to learn from examples as humans do. Therefore it is necessary, on the one hand, to have adequate emotional labels, which mostly derive from psychology and we discuss their power for HCI in Sect. 6.2.1. On the other hand, it is necessary to have convenient data sets at one's disposal, which can be used as training material. The second part in Sect. 6.2.2 is all about the data, its characteristics and the associated challenges that come with it. The preprocessing steps, such as feature extraction or selection, are presented in Sect. 6.2.3, followed by the concepts of classification in Sect. 6.2.4.

6.2.1   Psychological Emotion Models Leading to Labels

Over the last decades, psychologists developed different models to categorize human emotions. One of the first researchers dealing with this topic was Charles Darwin [18]. The varied categories can be divided in three classes: by name, categorical and (quasi)continuous.

The most famous concept to give the emotions names was developed by Paul Ekman [23]. He stated, that there exist six *Basic Emotions* universally recognized by all humans of different cultures: *anger, disgust, fear, happiness, sadness,* and *surprise*. These classes certainly are not sufficient enough to describe the emotions occurring in HCI.

Plutchik's wheel of emotions [57] is based on the idea of basic emotions, whereas he announced two more: *anticipation* and *trust* (see Fig. 6.1). On the one hand, this circumplex model is an emotion circle, so that similar emotions are located side by side and very different (bipolar) emotions on the opposite side. On the other hand, the model is like a color wheel, where the intensities of the different emotions are

Fig. 6.1 Circumplex and opened representation of Plutchik's emotion model: the eight bipolar emotions are arranged according to their similarity. The color saturation accentuates the intensities of the emotions, and the combinations of the basic emotions are written in between. Image adapted from www.puix.org

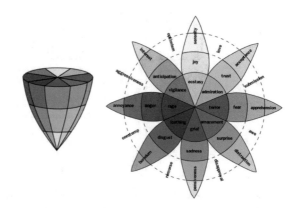

displayed by color saturation. He claimed, that all other emotions are combinations of these eight primary, genetic based emotions. Hence, he categorized the named emotions.

The *Geneva Emotion Wheel (GEW)* [63, 64] is also a categorical model. Version 2.0 consists of 40 emotions arranged in 20 families. Two axes (one for negative and positive valence and the other for low and high control) split the emotions in four separate groups. Also, a neutral class (or other emotion) is included in the center of the wheel.

The first dimensional approach was initiated by Wilhelm Wundt [81], who suggested that human emotions or feelings can be integrated in a three-dimensional space. The emotions differ in their quality (negative-positive), in their intensity (low-high) and in their tension (relaxed-tense). Based on this idea, James A. Russell and Albert Mehrabian [61] formed the three-dimensional concept of *Valence-Arousal-Dominance (VAD)*. These three continuous axes allow to arrange emotions more flexible and therefor more subjective. The first axis describes the valence of the emotion between negative and positive. The second represents the intensity of the felt emotion, whereas the third axis values the control you have over a specific situation. The so-called *Self-Assessment Manikin (SAM)* [8] allows persons to integrate their experienced emotion very easily in the VAD space (see Fig. 6.2). If possible, the self-rating is included in the recording of data sets.

Unfortunately, not all scenarios can be designed in the way, that the subjects designate their experienced emotion and, as mentioned before, not all data sets offer their emotional labels by design. Then it is necessary to label afterward, which means that some experts have to analyze the material several times in order to identify the emotional states. This procedure takes a lot of time and qualified personnel is needed. The only way to enhance the labeling process is to use good and handy tools, potentially take over some of the work. The *ATLAS* annotation tool offers these benefits [50]. It is a graphical tool (see Fig. 6.3), which allows to inspect and work with the raw data, but also with labels, features or even outputs of pre-trained classifier modules. There are several other annotation tools, but *ATLAS* is (as far as the authors know) the only one which includes active learning techniques, which

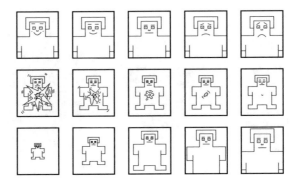

Fig. 6.2 The SAM used to rate the affective dimensions of valence (*top panel*), arousal (*middle panel*), and dominance (*bottom panel*). The subject can select the manikin in each line, which fits best the experienced emotion. Afterward, the values can be transferred in the three-dimensional VAD space

Fig. 6.3 Screen shot of the ATLAS annotation and analysis tool. The most important windows are shown (line-track window, video windows, main controls, label details window, tabular label view and the legend window)

means that the system utilizes already labeled parts to suggest labels for unseen sections. The annotator just needs to accept or change the recommendation, which is much less time-consuming than analyzing the whole data single-handedly.

6.2.2 Emotional Data Sets and Their Attributes

In the beginning of emotion recognition research, most data sets were recorded in a controlled environment, containing expressive acted emotions. In addition, they are restricted to one modality, like video stream (*Cohn-Kanade data base* [14]) or audio stream (*Berlin EMO-DB* [10]). Furthermore, the data sets are composed of short sequences without context. The achieved recognition rates on these data sets are promising [41, 67], but classifiers trained on this data perform extremely poor when applied in real-live scenarios. The reason is, as mentioned before, that the provided emotions are acted and exaggerated, whereas real-live conditions are influenced by changing light conditions, background noise and so on.

In recent years, several data sets have been published, trying to reduce the gap between acted and real-live scenarios. One idea is to record data in a so-called Wizard-of-Oz (WoZ) scenario [43], where an instructed expert controls an apparently smart system from a separate room, while the user interacts with it. The user is not aware of the expert and gets involved with the system, leading to real emotions. The expert needs to follow a restricted story book to induce predefined emotions at predefined points in time. This procedure leads to labels given by design [78]. The main drawback is that there is no control at any point (design, execution, analysis), whether or not the emotion induction was successful. An other idea is to arrange a more realistic scenario and don't limit the subjects in their behavior. This can be achieved by establishing a conversational situation between the subject and an avatar, leading preferably to a free, emotional colored interaction, as in [75]. But then it is not possible to assign labels previously, they have to be identified later. This process can be difficult and can lead to numerous problems (see Sect. 6.2.1).

In both cases, WoZ scenario and free interaction, the emotions occur rarely. Hence, the data bases are unbalanced with a majority of neutral states. This is a huge challenge for the classification process, as the problem that most of the data sets consist of many subjects, each with little samples. Since every person reacts slightly different, the best results can be achieved by treating each subject individually [7]. Therefor, plenty of samples are needed (see also Sect. 6.2.4).

Another aspect is the concept of multimodality in the data sets. Since each channel (audio, video) delivers other information about a user's condition, often at different points in time, the possibility to access both input streams can lead to much better classification results [69]. Nowadays, a third modality begins to play an important role in emotion recognition: the bio-physiological channel. Features extracted from sources such as heart rate or skin conductivity are not as controllable by the subject as facial expressions [42]. Despite that, the development of small and wearable measuring devices is flooding the market. In the near future, data sets will be published with other interesting modalities, originating from measuring tools such as time-of-flight cameras or gyroscopes.

6.2.3 Different Feature Types and Further Processing

To ensure proper classification results, it is necessary to extract adequate features, containing the required information to distinguish the annotated classes. The first step is the improvement of the raw signal, by applying filtering methods to reduce noise and detrend the signal, and normalization methods to accentuate contrasts.

Afterward, numerous features can be extracted. In the following, a fraction of well-known features is introduced, divided in video, audio and bio-physiological source.

6.2.3.1 Video Features

The feature extraction methods can be split in two groups: model-driven or data-driven. The model-driven approaches utilize the knowledge, that HCI involves humans. Therefor, faces and body parts appear in the video stream. Particular elements, such as eyes or the mouth, can be important for the emotion recognition.

Facial Action Coding System (FACS): FACS is a method to code facial expressions, depending on involved muscles. Paul Ekman and Wallace V. Friesen developed it in 1976 [24]. Each muscle movement is related to a so-called *Action Unit (AU)*, which offers the possibility to represent each facial expression as combination of AUs. For instance, AU1 indicates the raise of the inner eye brow. The six basic emotions (see Sect. 6.2.1) can be described by specified combinations of AUs. FACS was developed as technical manual for human experts to annotate the occurred expressions. The computer vision community researches automatic FACS identification algorithms, e.g. the *Computer Expression Recognition Toolbox (CERT)* [45].

The data-driven feature extraction methods are not limited to faces or body parts. Hence, they are used in a huge variety of classification tasks regarding pictures or video streams beyond HCI.

Optical Flow Estimation: This feature depends on the movement of pixels from one image to another. The result for each pixel is a vector $u = (u_1, u_2)$, containing the direction of the movement (for more information see [73]). These vectors can be transformed in a scalar value with $s_u = atan2(u_1, u_2)$ and they can be registered in a histogram, which can be used as input for a classifier. Because this methods calculates a feature out of two images, it is only applicable for video streams and not for single images.

Local Binary Patterns from Three Orthogonal Planes (LBP-TOP): The Local Binary Patterns (LBP) are features depending on the texture in an image (for more information see [53]). The 3×3 neighborhood of each pixel is considered to calculate a special value for the pixel, including the texture information of the whole region. These values are registered in a histogram and can be used as input for a classifier.

The LBP-TOP extends this idea by taking the time into account and can be applied on video data [83]. A video segment can be seen as voxel cuboid with three orthogonal planes: XY, XT and YT. Each direction leads to a LBP histogram, as mentioned before. The three histograms are concatenated to one, which can be used as input for a classifier.

6.2.3.2 Audio Features

Typically, audio features are extracted from short, overlapping windows, where the so-called *Hamming function* is applied. It is defined as: $w(k) = 0.54 + 0.46 \cdot \cos(\frac{2\pi k}{M})$, $k = -\frac{M}{2}, \dots, \frac{M}{2} - 1$, with window size M and sample index k. This filtering function prevents edge effects which can cause problems in later calculations.

Mel Frequency Cepstral Coefficients (MFCC): This representation is motivated by the way how humans hear. Sound waves stimulate specific cells in the cochlea, depending on their frequency and the perceived loudness is measured in this region by the number of impulses. The relation between frequency and perceived loudness is mapped on the so-called *Mel scale* (for referenced sound signals). The MFCC is a representation of the power spectrum in the frequency domain, mapped on the Mel scale [46]. They are utilized because of their compact ability to represent the important information in speech processing applications while retaining most of the phonetical significant acoustics.

Linear Predictive Coding (LPC): In LPC the speech sample s(t) is approximated by a linear combination of the past p samples [38]: $s(t) \approx a_1 s(t - 1) + \cdots + a_p s(t - p)$, where the coefficients $a_1, \dots, a_p$ are assumed to be constant in a short signal segment. This results in a p dimensional vector corresponding to a curve fitting the peaks of the short term log magnitude spectrum of a signal. As in the previous features the information is compressed, avoiding a transformation from time to frequency domain.

Voice Quality Features (VQ): Voice quality features are a group of features which are responsible for different speaking styles [47]. They all depend on the human vocal tract, the glottal flow and the lip radiation. These three parts are utilized to build a so-called Liljencrants-Fant model of the speech production process, which then leads to a variety of features. The parameters of the model are estimated from the scanned speech signal. One feature is the *amplitude level difference* $\Delta H_{1,2}$, which measures the offset between the first two harmonics of the narrowband glottal source spectrum H_1 and H_2.

6.2.3.3 Bio-Physiological Features

These statistical feature types are extracted from biological signals in humans, such as brain activity or heart rate. Unlike with the before mentioned sources, it is almost impossible to suppress or influence the bio-physiology behavior, which makes it so crucial for emotion recognition. To record these signals in time, special devices are needed. Some of them have to be adjusted precisely on the human body (electroencephalogram, electromyogram), others are much simpler and can be integrated in consumer devices such as mobile phones, smart watches or even in intelligent materials. Due to this big advantage, we focus just on these features, although the electroencephalogram and the electromyogram are of huge importance for emotion recognition.

Skin Conductance (SC): The German physiologist Emil Heinrich Du Bois-Rey-
mond was the first scientist who discovered the electrical conductivity of the
human skin in 1849. The skin conductance is highly related to the activities of
the sweat glands. If the sympathetic nervous system of a person is stimulated,
the perspiration increases and the resistivity of the skin descents. The so-called
electrodermal activity (EDA) can hardly be influenced and is one of the simplest
measurements that can be utilized for affective state recognition.

Blood Volume Pulse (BVP) and Heart Rate (HR): The BVP indicates the relative
amount of blood flowing through a blood vessel by a photoplethysmographic
sensor, normally fixed at one finger. It is depending on the heart beat volume and
on the vessel resistance. An infrared light is sent on/through the finger and the
reflection is quantified, whereat the intensity of light reaching the sensor is higher
when the blood flow is greater. At each heart beat, the BVP curve changes and
it is possible to measure the heart rate (number of peaks in one minute) and the
heart rate variability (variation in beat-to-beat interval).

Respiration (R): To investigate the respiration, a flexible sensor band can be placed
around the thorax and it measures the relative expansion/contraction of the chest.
The respiration rate (number of breaths taken in one minute) can provide information about the stress level of the human, also about eventually occurring panic or
asthma attacks. Nowadays it is possible to integrate respiratory rate detectors in
car seats and belts or even measure the respiration rate via phase detection radar
or acoustic signals.

6.2.3.4 Feature Post Processing

Typically, extracted features still have a sparse information density, which can be
improved by different methods, which either try to find the most relevant samples in
a data set or reduce the dimensionality in the feature space to the most informative
ones. By annihilating redundancies and focusing on the most relevant information
in the features, computation time can be saved and the errors rates can be reduced.

Instance Selection: This method "cleans" the training data set by discarding
extracted feature instances which have less or even confusing information for
the classifier by keeping only the most relevant ones. This can be done, for example, by training a weak classifier on the whole data set and utilize it afterward on
the training data set. Instances, where the classifier's decision and the ground truth
do not match are discarded from the training set, also instances where the initial
classifier has low certainty in the results [49]. Afterward, the reduced data set is
more separable and less noisy, resulting in a more accurate classifier. In the event
of large noisy data sets, the procedure can be repeated several times. To avoid the
problem of training and testing on the same subset of the data, the reclassification
can be executed as a cross-validation procedure.

Principal Component Analysis (PCA): This procedure is a very common, unsupervised statistical dimensionality reducing method, utilizing an orthogonal trans-

formation to map a data set of correlated variables into a set of values of linearly uncorrelated variables, the so-called principal components [40]. This transformation is defined in such a way, that the first principal component lies in the direction of the highest variance in the data, followed by the next one in orthogonal position, and so on.

Linear Discriminant Analysis (LDA): This method is closely related to PCA, but includes the label information of the data points [2]. The resulting transformation reduces the dimensionality, always being aware of the class memberships and tries to find a projection of the data points where the classes are separated as best as possible.

Forward Backward Selection: This procedure is a feature reduction process [32]. Initial, a classifier is trained on each feature separately. Then every possible remaining feature is added iteratively, the classifier trained and evaluated. In each step, the most significant feature is selected and added. This procedure is repeated until the classifier's performance does not increase any longer or even decreases. Now the backward step begins by testing each feature for its impact on the classification performance. The least significant feature is dropped and the reduced model is trained. This procedure is repeated until no more increase in the accuracy can be achieved. The remaining features are designated to be the most informative ones. Due to the selection process the selected features are not necessarily the global optimum. Additionally, it can be observed in the emotion recognition domain that the selected features strongly depend on the data set and the result is normally not intuitive [41].

6.2.4 Different Concepts for Classification

After the extraction of adequate features, the necessary post processing steps, and the agreement on labels associated with the occurring emotions, a classifier architecture can be build. As mentioned before, the machine learning concept is data driven and learns the best possible mapping of the feature instances on the emotional classes through training and adaptation. Afterward, it has the ability to classify new data samples depending on their similarity to already known ones.

Over the last decades, several classification concepts, models and architectures were developed and enhanced ever since. Also the idea to fuse available information on different levels gained influence. In the following, some well-known classifier architectures are described for example in Sect. 6.2.4.1 and analyzed towards their performance in HCI. Then, the different fusion levels are illustrated and their advantages are shown in Sect. 6.2.4.2. One absolute important aspect, concerning emotion recognition, is the existence of uncertainty. Experts, who label emotions, are often discordant and to not loose this information through voting, one can use fuzzy labels. Additionally, sensors can be noisy and to deal with this problem, each sensor gets a confidence value. How to deal with uncertainty and solve possible problems is described in Sect. 6.2.4.3. The last part of this chapter in Sect. 6.2.4.4 concerns the idea of classifier adaptation. During the system is operating, new data can be collected. A classifier, which has not the ability to adapt, has to be trained from scratch with the old plus the new data. An adaptive classifier is able to include the new information, without retraining.

6.2.4.1 Classifier Architectures and Their Potential for HCI

In the following, some elementary different classification approaches are presented. *Support vector machines* belong to the supervised two-class classifiers, but can easily be adapted to more than two-class problems. They are one of the most utilized classification concepts, because of their huge adjustability. The *hidden Markov models* utilize an underlying *Markov process* to deal with time dependencies and are therefore often used for feature sequences. They can also easily deal with uncertainty in the labels (see Sect. 6.2.4.3). The third idea regards the existence of large and most notably complex emotional data sets, which are used by *Deep neuronal networks* utilizing cascades of different layers to learn different representations corresponding to different abstraction levels.

Support Vector Maschine (SVM): A SVM optimizes a hyperplane between two classes in the way that the margins between the classes and the hyperplane is maximized [15]. This hyperplane is defined by selecting informative data points from the input set, so-called support vectors, which are as close as possible to the hyperplane (see Fig. 6.4). New examples are then compared to this hyperplane and predicted to belong to a category based on which side of the plane they fall on. SVMs can efficiently perform a non-linear classification using what is called the kernel trick, implicitly mapping their inputs into higher-dimensional feature spaces. Non linear classification is important in the HCI domain, due to the non linear and overlapping structure of typical HCI data. Furthermore, they can also be utilized to solve multi-class problems by implementing so-called one-versus-one SVMs, where all pairs of possible two-class SVMs are trained and the decision is made by voting. There exist many extensions of the basic SVM algorithms, like fuzzy input fuzzy output SVM's (F^2SVM) [66, 74] which can deal with uncertainty of the given labels and can also emit uncertainty information. Also

Fig. 6.4 Example of a SVM: hyperplane *h* separates the *classes 1* and *2* with the greatest possible margin. The support vectors on the border are marked with thicker edges

adaptive SVM (aSVM) [82] and transductive SVM (tSVM) [39] are useful in dealing with large partially labeled data sets.

Hidden Markov Model (HMM): Compared to SVM learning, where each feature vector is computed independently, HMMs are utilized for the classification of sequential data. The time window of a feature is often shorter than the experienced emotion, hence emotions can be composed of several time segments and build sequences of features. Therefore, HMM gain more and more importance for HCI [58]. A HMM is a statistical model, which is composed of two random processes. The first is a Markov chain consisting of a fixed number of states and assigned state transition probabilities. This process is hidden and only the output, the second random process is visible. Each of the states has a probability distribution over all possible output tokens (Fig. 6.5). Therefore the sequence of tokens, generated by a HMM gives some information about the sequence of states. Usually one single HMM is trained for each class and the unknown sequence is presented to each of them. The HMM with the highest logarithmic likelihood output of the possibility, that the presented feature sequence was produced by it, designates the class.

Deep Neuronal Network (DNN): Due to the increasing computation power and availability of large emotional data sets, deep learning became more and more popular in recent years. One manifestation of deep learning are DNNs [72] which are inspired by the way our human brain is build, where also large neuronal structures of complex layered structures can be found. The idea is to represent different abstraction levels of the input data in the different layers and model eventual non-linear relationships between them. In lower layers basic structures of an object are learned, whereas in upper layers this structures are more and more merged to complex compositional models up to the model of the final class. The necessary structures are learned by the network itself. DNNs are typically designed but not limited to feed forward networks. *Convolutional deep neural networks (CNNs)* [48] have been applied in computer vision and adapted to vision based emotion recognition, to acoustic modeling for automatic speech recognition and to speech based emotion recognition. Due to the complex and large structure of DNNs, a lot of training material is needed which leads to computational expensive training.

6.2.4.2   The Concept of Fusion on Different Levels

Towards the multi-modal character of human emotion expressions, building a proper fusion architecture is crucial. The different data sources (e.g. camera, microphone) record information at the same time, but not necessarily with the same content. It is possible, that one sensor is not reacting on the user's emotional changes, because of a malfunction or because of situative reasons, e.g. the user is not looking towards the camera, not speaking or not touching a sensitive surface. Also not every emotional behavior can be detected in every modality. For example a short smile does not

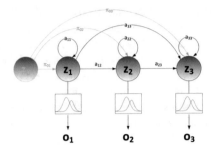

Fig. 6.5 Example of a HMM: The Markov Chain consists of three states (Z_1-Z_3) and one initial state (Z_0). The corresponding state transition probabilities are assigned to the edges. The second random process concerns the output, which is often defined by *Gaussian mixture models*

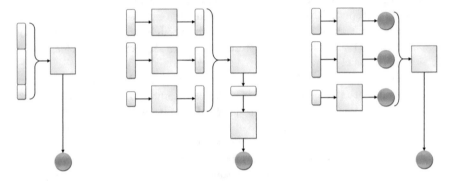

Fig. 6.6 Schematic drawing of fusion architectures. From *left* to *right*: Early, midlevel and late fusion. The most *left* boxes depict the feature instances going into the classifier modules (next column of *square boxes*). In the midlevel fusion, internal data representations evolve from the classifiers and are passed to the next level. The *circles* stand for the crisp classifier decisions

necessarily cause a change in the skin conductance. The fusion techniques have to deal with this problems [22, 65], accordingly to the existing reason. Figure 6.6 gives an overview of the fusion architectures presented in the following.

The most simple fusion architecture is the early fusion, where the feature vectors are just concatenated and treated as one high dimensional vector. This method enforces feature vectors which specify the same period of time. Since in HCI not all features are based on the same period in time, it is impossible to combine the underlying modalities. For example, audio and video features cover milliseconds up to a seconds, whereas changes in bio-physiological features, as heart rate or respiration, can only be detected in much longer windows up to 30 s length [77].

Late fusion assumes an individual classifier for each feature, giving an emotion class prediction. These predictions are then simply voted or melted to a final agreement. This can be very complex like a *Markov fusion network (MFN)*, which is a probabilistic model for multi-modal and temporal fusion introduced in [28, 29]. These so-called ensemble methods or mixture of experts, typical consist of many similar independent weak classifiers, while a final strong decision is computed out of the divergent weak single decisions. Empirically, ensembles tend to yield better results when there is a significant diversity among the models. In special cases it is helpful to use different classifier architectures as weak classifiers in order to increase diversity. Unfortunately, information might get lost on the way to the decision, due to the strong compression to just class labels in the step before. In order to increase the efficiency of late fusion it is desirable to use additional uncertainty information from the input level and use the fuzzy labels for the fusion step.

As a hybrid architecture, mid level fusion covers the advantages of both methods by simultaneously dropping the disadvantages. Mid level fusion consists of several classifiers and fusion steps which combine results at different steps in time, when they are meaningful or intuitive. For example, similar features, concerning the underlying period in time, can be combined in a relatively early stage (see early fusion), whereas different features have to meet late (as in the late fusion). Otherwise, the time scaling problem can be reduced by adding temporal fusion modules to the architecture. Additionally, the classes of a inner fusion module do not necessarily need do be the same ones utilized for the final output. For example, a smile detector can combine several video features and is then utilized as input for a later fusion step, which maps to the big six emotions.

6.2.4.3   Treatment of Uncertainty

Having the goal in mind to build realistic interaction systems with the ability to handle the user's emotional behavior, dealing with uncertainty is crucial. On the one hand, the ground truth of training data is uncertain due to the complexity of emotions in the HCI domain. Hence, the classifiers have to deal with fuzziness or uncertainty on their input side. It is not realistic, that every emotional state of the user is 100 % clear. Even though the emotional states are labeled by experts, it can happen that they do not agree on the labels. On the other hand, the classification of real-life emotions is difficult and the reached classification rates are not always convincing. So it can happen that a classifier makes a decision for one class just because the result was a little bit higher than for the second class. Hence, if the result is going to be passed to a fusion architecture, the fuzzy output can be the better one to work with. To sum up, it is necessary to map uncertainty from the input up to the classifiers output and give the system information about the reliability of the given emotional state prediction at any point during the classification process.

One approach to deal with uncertainty is formulated by Arthur P. Dempster and Glenn Shafer known as the *Dempster-Shafer theory (DST)* [70]. The theory allows the combination of evidence from multiple sources and results in a value for the degree of belief, which is described by a belief function taking into account all the available information. Dempster introduced methods to combine degrees of belief from items of multiple sources and his student Shafer developed a method to obtain degrees of belief for one question, using probabilities for a related question. This means, each classifier can be treated as an expert giving an answer to a question.

Using Dempster's rule of combining [19], these degrees of belief can be combined to solve the one question about the classification label. The output consists of two related values: the probability for one class and the belief in it. The system can then decide, how to treat this information.

6.2.4.4 Individual Adaptivity of the Classifier to One User

The classification and the fusion process have in common that a huge amount of date is needed to build adequate models, which are then able to describe the human emotional behavior. These large data sets are mostly collected from a group of humans, where the data of each person forms just a small part of the whole data set. Empirical knowledge shows instead, that emotion recognition performs better if each human's behavior is modeled individually. Hence, a huge data set for each person is needed. Another approach to deal with the gap between weak performing inter-individual recognition and well performing individual recognition, is to utilize adaptive classifiers and adaptive fusion architectures. In this case, a initial structure is trained on a large data set in order to learn general human emotional behavior. Then, this initial structure is adapted to one user individually by using training material collected in a short recording session. It is also possible to further adjust the classifier during the interaction between the user and the system with additional data, gathered during the system's inquire phases (see Sect. 6.3.2.2). Model adaption is only reasonable, when it works without retraining the whole model with the initial data set and the new recorded samples. It must be possible to include the new information easily and fast in the existing classifier. When working with SVMs for example, adaptation can be achieved by comparing the new data samples only with the existing support vectors and decide afterward if one of the new points is going to be a new support vector and replaces an old one. This is a much faster procedure than to learn a new and probably similar hyperplane with the whole data set.

6.3 Concept of Emotion-Adaptive *Companion*-Systems

In this section we explain the possible components of a *Companion*-System along with our component stack. Then, we motivate and explain our approach for systematically emotion-related adaption of dedicated components.

6.3.1 System Architecture

The different layers of our system architecture (see Fig. 6.7) can be mapped to the arch model from Gram and Cockton [31]. The presented components allow the system to operate as follows (cf. [36]). The highest level in Fig. 6.7 contains the

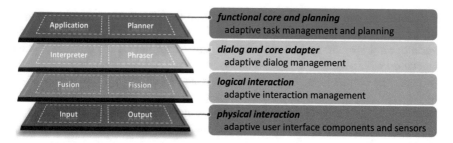

Fig. 6.7 Adaptive architectural component stack of a *Companion*-System. All presented eight component clusters (i.e. Application, Planning, Interpreter, Phraser, Fusion, Fission, Input, and Output) from four different layers are able to react to a recognized change in the user's emotional state

functional core of an application (e.g. a media player with functions for decoding media and managing the media library). This level can also contain a component for AI *planning* of assistance. If the user needs to solve a complex problem, the planner is in charge to infer a sequence of tasks which lead to a solution of the problem. The level of the *dialog and core adapter* serves as modality-independent interface to the core's functionality on the one hand, and is able to communicate the content of each necessary task step, on the other hand. The *Phraser* has to provide a user-individual dialog structure for the system's intended output. The *interpreter* has to interpret user inputs and has to decide about which of the core's functions have to be called, or which of the planner's tasks are fulfilled.

The next level represents the *logical interaction* layer. The *fission* component refines the dialog management's abstract output and reasons about the modality-specific output. Thereby, it decides about the use of the later applied device components and sensors [36]. The second component on this level is the *fusion* component. This component is able to fuse multimodal inputs and derive semantic meaning, which serves as input for the dialog management's interpreter. The lowest layer for *physical interaction* comprises all available device components for *input* and *output*. Output components are determined by the fission component and render the dialog manager's intended output in a multimodal manner. In the same way, input sensors are provided with specific grammars or configuration data, which allow to recognize possible implicit (e.g. user's emotional state or location) as well as explicit user inputs (e.g. user's speech, touch, or gesture input).

6.3.2   Identifying the Cause of Emotions for Proper Reactions

Possibilities for adaptation on each of these different levels rises the question of: When shall what component on which level perform an emotion-related adaptation? It is obvious, that performing adaptations on every level whenever possible do not lead to the desired result in terms of emotion-reflective behavior. As an example, if a user does not like the current layout of the system's user interface (UI) when working on a specific $task_a$, it is better to only adapt the UI for that task via an adaptation within the fission process on the logical interaction layer, than to provide the user with another $task_b$, by performing an adapting the current task sequence via the planner component on the planning layer (cf. Fig. 6.7). Therefore, we aim on the following. If necessary, a successful identification of the cause of a sudden change in emotion shall allow a systematic adaptation in only one single emotion-related component cluster.

The success in doing so, mainly depends on the ability to identify the emotion-related object which can be most probably assumed to be the cause of a sensed, but unwanted emotional effect. Therefore, we advise to keep track of each component's actions. See Fig. 6.8 for an exemplary sequence of system outputs and inputs with layer-specific events over a certain period of time. These component specific action events are kept in memory and form a hierarchical, time-based event history, the so-called *interaction history* (see Fig. 6.9). Each item in the history represents a 5-tuple ⟨*time, component, trigger, action, expectedAffectiveReaction*⟩, a so-called *action item*. Where *component* represents one of the system's components that performed a specific *action* at a known time stamp *time* (e.g. UNIX time), activated by a given *trigger*. Since the system is aware of its actions and the quality of its own output [34], it can even expect emotional *reactions* in some cases. For example, if the system is aware that it cannot respect privacy issues in a specific UI setting [34, 62] (e.g. if a password prompt can only be realized via a device combination (speaker/microphone) that cannot enforce a desired privacy policy), or if

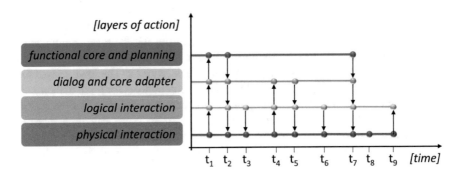

Fig. 6.8 Layers of action events over time. In the output direction (*top down*) actions on higher levels usually cause actions on lower levels and vice versa on the input direction (*bottom up*). Actions can not only originate from user inputs, but from any layer, caused by an adaptation to a change in the current context of use. The temporal-marked events are kept in memory as so-called *action items*. They allow a later reasoning about the possible mapping of an action as cause to a possible occurred emotional effect. For simplification, actions with causal relationship are printed at the same point in time, though implementations of this concept have to deal with offset time shifts. Another simplification is used on the physical interaction layer. Events from different interaction concepts (a.k.a. modalities) are visually summarized and printed as one single node, if they occur at the same time

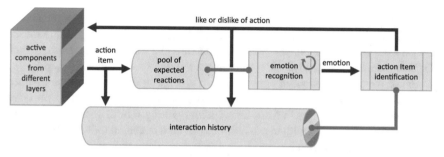

Fig. 6.9 The process of emotion-based action rating. Active components form diverse layers provide their action items. The items are stored in the interaction history for later use. Expected affective reactions are summarized to configure the ongoing emotion recognition. Components are updated if a user likes or dislikes a particular action

an unpleasant information has to be announced (e.g. the warning of a traffic jam ahead). This last sub-item is optional, because not every action item can provide such information about an expected affective reaction. If an action item provides such information, it can be used as useful hint to configure and support the process of emotion recognition (see Fig. 6.9), and can additionally be used to support the identification of the emotion-related object.

Whenever a shift of a user's emotion is detected, the system inspects its interaction history to identify the possible actions that took place right before the start of the sensed change in emotions. Thereby we can distinguish different cases for identification as described below.

6.3.2.1 One Possible Reason

In the best case, the inspection of the interaction history reveals only one possible item which could be identified as emotion-related object. As example, if the user rotates his smartphone the layout may switch from portrait to landscape, and this may result in a cropped view of the UI where some UI elements can not be displayed without the need to scroll. The user does not like the adapted layout with the scrollable UI. Such an action could be saved in the interaction history as action item 5-tuple ⟨ *1430923042637, phone.display, enterLandscapeMode, adaptationOfLayout, negativeReaction* ⟩ (cf. action at t_8 in Fig. 6.8). The referenced component's action from the fetched action item can be identified as the emotion-related object.

6.3.2.2 Multiple Possible Reasons

As mentioned before, the CoU can lead to a change in any of the architectural layers. In an exemplary situation, the output is transferred from one screen to another because of a moving user, as described in [36]. In such a case, the decision about the output takes place in the interaction management's fission component (cf. action at t_6 in Fig. 6.8). In the given example, this initial action causes a follow-up action on the lowest level with the involved output components. The former screen has to switch to idle mode, and the newly addressed one has to render the designated output.

Based on our approach, in such a case, a change in emotions can be mapped to multiple possible reasons. In the given example, the interaction history provides three possible action items that are linked with t_6 in Fig. 6.8.[1] The next paragraphs describe strategies how multiple possibilities can be reduced, in order to identify the one emotion-related reason.

Identification Based on Expectation: As motivated above, action items can contain an optional sub-item that provides information about an expected affective reaction. Such information can be used to disambiguate possible emotion-related objects and decrease the number of possible reasons. Items, for which the expected reaction equals the sensed affect can be attributed with a higher evidence for being the emotion-related object, than those with none or other expected reactions.

Reduction Based on Causality and Dialog History: Since the layers of action depend on each other, it is most likely that action items depend on a causal relationship if they occurred at (almost) the same time. In such cases, causal dependencies can be inferred by matching the *trigger* and *action* markers from each involved item as stored in the interaction history. For each causal chain (or hierarchy) that can be built up based on the equality of the action and trigger markers, we try to refuse subordinate action items based on the history showing that they are not involved.

In that regard, a history of uninvolvement can arise from similar former actions, at which no negative affect was recognized. If, for example, a media application continues in playing a playlist and starts a new song, this could cause actions on all lower layers (cf. t_2 in Fig. 6.8). But if the user was pleased with the system's former decisions when relaxing on the sofa while listening to the past songs, and the CoU has not changed (despite the song), we can rule out the actions on the lower levels (use of the same: dialog strategy, logical interaction, and speaker settings on the physical level) for being the cause of the emotion. So in that case, the newly played song might be the negative emotion-related object.

User Clarification: If the aforementioned strategies are not able to cut down the possibilities to one action item, a *Companion*-System is able to autonomously generate a new dialog at runtime and ask the user for clarification, as described in [37]. The content for such a template-based selection can be derived from the remaining action items' *component* and *action* data. Thereby, each selection item refers to one possible emotion-related object.

---

[1] In the referenced Figure, the two action items from the both displays are visually summarized and printed as one single node.

6.3.2.3 Summary

The presented approach is able to support emotion recognition at runtime by providing hints about expected user reactions. Further, it allows to map sensed emotions to interaction related objects. Thereby a multi-layer interaction history is used to identify an emotion-related object as well as to gather evidence for the correct decisions of the system (cf. Fig. 6.9). If the *Companion*-System cannot identify the referenced object by itself (with sufficient certainty), a dialog with the user can be initiated to clarify the cause of an unwanted user emotion. The so-gained knowledge about the system's emotion-related action items allow the system to perform layer-specific adaptations, which is not supported by other concepts, to the best of our knowledge. In the following we describe possible emotion-related adaptation strategies with focus on the two layers of logical and physical interaction.

6.4 Examples of Emotion-Based System Behavior

The interaction of a user with a *Companion*-System is realized via the logical and physical interaction layers (cf. Fig. 6.7). Thus, in order to perform a meaningful adaptation within these layers, the cause of change in the affective state of the user must lie within these layers as well. Overall, the goal of emotional adaptivity in the interaction layer is to decrease negative affect (or valence) and thus prevent dialog interruptions and avoid dissatisfaction. Depending on whether the input or output of the system is concerned, the causes of emotional user reactions, the reactions themselves, as well as the adaptions undertaken by the system may differ largely. To give some examples for emotional adaption, we will use the scenario of car driving. On the one hand, this scenario allows multiple different interactions with technical systems, on the other hand, the environment can easily be equipped with diverse sensors for emotion recognition. In the following sections, concrete examples of interaction situations with emotional adaptivity in the scenario are elucidated.

6.4.1 Physical Interaction Layer: Automatic Volume Control

Most of today's cars offer a built-in speed dependent volume control (a.k.a. SDVC, SCV, GAL or GALA). This feature adjusts the audio system's volume to the car's current speed. In that way the passengers do not have to adjust the volume manually, as driving noise increases with speed. In cases, where speed decreases abruptly, the volume may drop in a noticeable manner (cf. transitions of actions for automatic volume adaptation from t_7 to t_8, Fig. 6.8). Some drivers may not manage to fine-tune the settings for this adaptation process. Hence, it is possible that the automatic volume controller's adjustment decisions do not match the driver's opinion in some situations. A negative emotional expression can be detected by using a camera looking towards the driver's face. Short term features like LBP-TOP and FACS can be captured from this stream. Due to possible changes in the surrounding noise level, the audio information might not be used as input. Since the driver is in continuous contact with the steering wheel, it is possible to get additional information about the SC, which typically is responsible for immediate changes towards negative feelings. Decision level fusion is able to combine the facial and SC information to a final estimation of appearance of a negative emotional event.

A time-based look-up in the interaction history reveals the automatic volume adaptation as cause of the negative emotion (cf. process in Fig. 6.9). This finding is passed to the action item's linked active component, the automatic volume controller. Based on that feedback, the volume can be re-adjusted.

6.4.2 Automatic Modality Selection on the Logical Interaction Layer

In this section we focus on a *Companion*-System's fission component which has to decide about the mapping from a logical interaction concept to its physical representation (cf. Fig. 6.7), i. e. the fission reasons about the use of different modalities. This reasoning process can originate from actions within the two highest layers, but it can also be triggered from changes in the CoU, as described in the example below.

The driver is a commuter who regularly takes the same route from his home to the office and back. Despite the fact that he knows his route, he activates the navigation system on each trip. He does not want to take the risk of getting stuck, because of traffic jams. After several equal tracks, the *Companion*-System gained enough evidence and assumes, that the driver is familiar with these everyday routes, and would be happy if the system would suppress its ongoing verbal announcements for the already known turn-by-turn navigation.

The fission component decides on the logical level, that the traffic warnings shall still be presented via the graphical user interface (GUI) plus a verbal text-to-speech (TTS) output on the physical layer. The turn-by-turn announcements, in contrast, shall be solely presented on the GUI without any additional acoustic output as long as the driver stays on the well-known route. This adaptation to the CoU is represented by the transition from t_2 to t_3 (cf. Fig. 6.8). At the first turn without the accustomed acoustic announcement, the system detects slight confusion in the driver's affective expressions. The sensed expression can be dated to t_3.

Weak confusion is mostly expressed in slight changes of facial expressions, detectable by FACS based recognition or short term optical flow extraction. Despite the fact that verbal or bio-physiological reactions are not expected, they can also be monitored in cases where the driver is very expressive. A late or midlevel fusion method is able to get the correct estimation, even if there is only an evidence in one of the input modalities.

To identify the emotion-related object, the system starts its analysis on the physical interaction layer. Since the driver did not complain about the turn-by-turn announcements on the GUI in the past, the system gained evidence that the user liked the use of this device for that particular purpose (cf. Fig. 6.9). That is why the fission component can be identified as emotion-related object via its causal relationship (see t_3 in Fig. 6.8). Based on that new "dislike", the *Companion*-System might be able to discard its assumption, that the user would prefer the visual channel, and thus could continue using the former variant with multimodal output.

6.4.3   Multiple Layers: The Announcement of Traffic Warnings

Imagine a situation in the car, in which the navigation system wants to announce the current traffic warnings (cf. action at t_2 on the *functional core and planning* layer, Fig. 6.8). The respective *action item* is kept in memory as part of the interaction history. The navigation system knows by implementation, that warnings about upcoming traffic jams may cause negative emotions. That is why this action item offers optional information about an expected negative emotion as its designated sub-item (cf. Sect. 6.3.2). Invoked by the navigation system, the dialog management applies its only strategy for such a case; it prepares the output with a list of all identified warnings. Thereafter, the fission component gets activated with this modality-independent list of warnings on the logical interaction layer. Next, on the physical layer, each warning could be presented via different output concepts, like TTS or GUI. In this simplified example, the fission has to decide between the following three output options: GUI, TTS, or GUI+TTS. Based on knowledge about the actual CoU, the system ranks the options and decides that using GUI+TTS is definitely the best way to offer the proposed warnings. This new output supersedes the former output. The music from the radio task gets stopped. The linked meta-information about singer and song on the display give way to display the traffic warnings, and the speakers are used for TTS output.

Right after the aural announcement catches the driver's attention, the system detects a negative emotion and dates its origin at t_2. Again the video stream and short term bio-physiologic data are suitable to get knowledge about this emotional change. In case of verbal reactions of the driver, the audio information is also useful as input of the fusion architecture. In case of uncertainty about the detected emotion, the information about the expected reaction can be used as additional classification input. Due to the temporal character of the driver's reactions and the meta information about the expected behavior, a MFN is useful as fusion method here.

In this example, a chain of causal dependencies leads to the identification of four possible action items, one on each layer. Since the driver did not complain about the use of display and speaker in the past, we have evidence to refuse the possible emotion-related object on the physical layer, as described in Sect. 6.3.2.2. In this case, the process for emotion-related action item identification can succeed without further user assistance, because only one of the three remaining action items comes with an expectation for negative emotions. Based on the strategy described in Sect. 6.3.2.2, the navigation system's action item gets the highest evidence for being the emotion-related object. Since the sensed emotion was expected for this action, the system does not have to adapt its behavior.

6.4.4   Interaction Input Layer: Navigation Misunderstanding

A typical task in a car driving scenario is the specification of a destination, for which modern cars offer a speech input to avoid distracting text entries. To make sure the speech recognition was correct, often speech input has to be explicitly confirmed and misunderstandings require explicit user inputs. With an emotion aware system, these cumbersome dialog steps could be easily avoided, as no emotional reaction implicitly confirms the previous input. Explicit inquiries are only necessary and automatically triggered, when a negative emotion is detected, as described in the following.

Imagine the situation where the user says "Newark" to enter a destination. The speech recognizer misunderstands "New York" as the most probable input and forwards this to the upper layers in the component stack (cf. actions at t_1, Fig. 6.8). As soon as the assumed (but wrong) destination is displayed on the navigation display (cf. t_2, Fig. 6.8), the user would show a negative emotion in the form of confusion and/or anger.

Anger mostly causes facial reactions of the eyebrows which can be detected utilizing FACS. Additionally, a driver's verbal reaction like "oh" is possible and a few MFCC or LPC instances can be extracted. In case of interacting with the system while driving, the driver touches the steering wheel and therefore bio-physiological information is available (HR, SC). Because of the unclear situation which sensor is currently delivering information, a MFN is suitable as fusion architecture. In this case, anger appears suddenly and the classification of confusion and anger takes place at the same time. This can be realized by either two separate classifiers, resulting in a multi-label answer or via one single network, specialized on confusion driven anger.

As previous speech inputs have worked as expected, the reason for the emotion can be assigned to the physical interaction layer, i.e. the speech sensor. The system is now aware of the misunderstanding and displays the possible alternatives (in order of their recognition confidences) or provides an option to repeat the input. As the correct destination "Newark" is within the list of alternatives, the user can select it and start navigation.

6.4.5   Interaction Input Layer: Seat Belt Reminder False Positive

Let's assume the family puts a heavy object (e.g. a heavy travel bag) on one of the rear seats. This could be interpreted by the occupant detection system as a person sitting there, although that is not the case (hence the term false positive). Thus, the seat belt reminder would issue an event at t_1, leading to an annoying warning signal at t_2 in Fig. 6.8. This might cause an emotional reaction of surprise at first and even anger after a while.

Surprise often causes strong facial reactions which can be detected utilizing any short term feature described in Sect. 6.2.3. Additionally, a driver's verbal reaction like "oh" is conceivable and useful. However, VQ features can not be extracted from such short utterances, so MFCC or LPC need to be used. Bio-physiological information are only suitable when the driver touches the steering wheel for some time, giving the system time to calibrate itself. In the beginning of this scenario, this is not necessarily the case. Due to the strong presumption of missing inputs a MFN is suitable as fusion architecture. The change towards annoying anger needs some time, therefore we can assume that the system is now calibrated and bio-physiological information about HR and SC are available, also respiration is suitable due to the slow change towards anger. Overall, a selection of the full bag of features can be used to drive a probabilistic mid level fusion architecture, to make the system aware about the user's anger.

As the interaction history contains only actions (of the other layers, e.g. creating a warning dialog) that do not allow any variations in handling such a warning, the only remaining possibility can be assigned to the initial input of the seat occupancy system to be a false positive. Therefore, the system could ask the user to temporally disable the warning message when the seat is not occupied by a human. Because the warning is safety relevant, we would not recommend a silent automatic deactivation.

6.4.6   Interaction Input Layer: Media Control False Negative

The rear seat entertainment allows for convenient multimodal interaction to control the play state of media by using gaze and voice commands in combination. For example, if the system detects the gaze on the media screen, voice commands can be used to play/stop/pause the current movie or audio track. Thus, cumbersome touch interactions can be avoided, which is especially helpful for kids having difficulties to reach the monitor when the seatbelt is fastened. Let's assume a situation, where the child is listening to some music and is watching the scenery passing the side window. Now he/she wants to stop the music in order to ask the fellow travelers for the name of the region they are crossing. Therefore, it gazes at the media screen and utters the "stop" command. In such a situation it can easily happen that the voice command was too quiet to trigger a speech input in such a noisy environment. As the speech sensor does not react upon an actual event, we call this a false negative. Because nothing happens, the child responds with an emotion of impatience that could be detected by a cheep camera placed in the media screens frame. Additionally, acoustic information, which is too silent for speech recognition, can be clear enough to get some affective state information from it. Furthermore, a reaction like "hmm" can be considered as reaction of impatient. Bio-physiological data can not be used in this case, because there is no contact to a corresponding sensor. In our opinion, impatience needs up to several seconds to raise, so features driven from windows up to a second or two can be used. This includes almost all video and audio features except VQ due to the shortness of the utterance. It is not advisable to use all features as they are, because of computation time, therefore the dimensionality of video features should be reduced by using PCA or LDA methods. In this special case, we suggest an early fusion to recognize a continuous impatient value, which is easy to perform when using similar window lengths for the audio and video features.

In this particular situation of complementary input according to the CARE properties (cf. [16]), only the speech modality (input device component) is responsible for the false negative and the gaze sensor still provided some input. Such way, evidence for a possible false negative in the input layer is available, in addition to the interaction history as shown in t_8 of Fig. 6.8.

As an immediate reaction, the system apologizes for the failure and kindly requests to repeat the input that is detected correctly now. To avoid such false negatives in the future, the system could increase the sensitivity of the affected input device, i.e. the speech sensor in this example. Despite this obvious adaption that bears the risk of inducing other errors, there exists another possibility for the given multimodal case. As pointed out in [68], when the typical temporal relation of the involved modalities is known (maybe even in a user individual way), such information can be used in the fusion of the logical interaction layer to detect and resolve such false negatives in the first place.

6.5   Discussion and Future Work

After an comprehensive discussion of theories of emotion and the state of the art in emotion recognition we presented an architectural layer stack of *Companion*-Systems that comprise specialized components for interaction and dialog management, as well as for AI planning. Using this layered model we proposed a specific interaction history and a corresponding process that can be utilized to solve the problem of assigning a detected emotion to a specific component as the root cause for the emotion. Having identified which component of a system is responsible for the detected emotion, adequate adaptations can be initiated to reduce any negative affect. Positive affects can be used to identify the system's applied strategies, which are favored by the user. Future work can include such hints when reasoning about the most adequate action, dialog strategy or form of output. As described in several examples of an in-car scenario, the approach allows adequate reactions of the interaction components to detected changes in the emotional state of a user.

The limits of feasibility for the proposed process need further exploration. Until it is implemented into a test-bed application suitable for extensive user-tests, it remains unclear if it will work as expected, or to what extent heuristics will have to be incorporated to guide the reasoning process. We expect problems when a strategy uses *causality and historic evidence* (Sect. 6.3.2.2), as long as the interaction history is rather small. Additionally, strategies to resolve unclear reasoning results and applied adaption strategies (some of which are exemplified in Sect. 6.4) must be further investigated.

Another topic open to discussion is the use of a dedicated model of emotions for HCI, as the classical psychological models (cf. Sect. 6.2.1) often appear either too generic or lack adequate categories needed in HCI. Though building such a model needs further and extensive research, it is of utmost importance to drive the development of emotion recognition components which are suitable, sensitive and robust enough to be used in HCI as described here.

Acknowledgments This work was supported by the Transregional Collaborative Research Center SFB/TRR 62 "Companion-Technology for Cognitive Technical Systems", which is funded by the German Research Foundation (DFG).

References

1. Akgun M, Cagiltay K, Zeyrek D (2010) The effect of apologetic error messages and mood states on computer users' self-appraisal of performance. J Pragmat 42(9):2430–2448. doi:10.1016/j.pragma.2009.12.011 (how people talk to Robots and Computers)
2. Altman EI, Marco G, Varetto F (1994) Corporate distress diagnosis: comparisons using linear discriminant analysis and neural networks (the italian experience). J Bank Financ 18(3):505–529
3. Anderson K, André E, Baur T, Bernardini S, Chollet M, Chryssafidou E, Damian I, Ennis C, Egges A, Gebhard P, Jones H, Ochs M, Pelachaud C, Porayska-Pomsta K, Rizzo P, Sabouret N (2013) The tardis framework: Intelligent virtual agents for social coaching in job interviews. In: Reidsma D, Katayose H, Nijholt A (eds) Advances in computer entertainment. Lecture notes in computer science, vol 8253, Springer, Berlin, pp 476–491. doi:10.1007/978-3-319-03161-3_35
4. Atkinson A, Dittrich W, Gemmell A, Young A (2004) Emotion perception from dynamic and static body expressions in point-light and full-light displays. Perception 33(6):717–746. doi:10.1068/p5096
5. Bastide R, Palanque P (1999) A visual and formal glue between application and interaction. J Vis Lang Comput 10(5):481–507. doi:10.1006/jvlc.1999.0127
6. Becker-Asano C, Ishiguro H (2011) Evaluating facial displays of emotion for the android robot geminoid f. In: 2011 IEEE workshop on affective computational intelligence (WACI), pp 1–8. doi:10.1109/WACI.2011.5953147
7. Böck R, Gluge S, Wendemuth A, Limbrecht K, Walter S, Hrabal D, Traue HC (2012) Intraindividual and interindividual multimodal emotion analyses in human-machine-interaction. In: 2012 IEEE international multi-disciplinary conference on cognitive methods in situation awareness and decision support (CogSIMA), pp 59–64
8. Bradley MM, Lang PJ (1994) Measuring emotion: the self-assessment manikin and the semantic differential. J Behav Ther Exp Psychiatry 25(1):49–59
9. Brave S, Nass C (2003) Emotion in Human-computer Interaction. In: Jacko JA, Sears A (eds) The human-computer interaction handbook, L. Erlbaum Associates Inc., Hillsdale, NJ, USA, pp 81–96
10. Burkhardt F, Paeschke A, Rolfes M, Sendlmeier W, Weiss B (2005) A database of german emotional speech. In: Proceedings of interspeech, Lissabon, pp 1517–1520
11. Campbell N, Kashioka H, Ohara R (2005) No laughing matter. In: 9th European conference on speech communication and technology, INTERSPEECH 2005—Eurospeech, Lisbon, pp 465–468, 4–8 September 2005

12. Castellano G, Villalba S, Camurri A (2007) Recognising human emotions from body movement and gesture dynamics. In: Paiva A, Prada R, Picard RW (eds) Affective computing and intelligent interaction, vol 4738, Lecture notes in computer science, Springer, Berlin, pp 71–82

13. Cohen I, Sebe N, Garg A, Chen LS, Huang TS (2003) Facial expression recognition from video sequences: temporal and static modeling. Comput Vis Image Underst 91(1–2):160–187. doi:10.1016/S1077-3142(03)00081-X,specialIssueonFaceRecognition

14. Cohn JF, Kanade T, Tian Y (2000) Comprehensive database for facial expression analysis. In: Proceedings of fourth IEEE international conference on automatic face and gesture recognition, pp 46–53

15. Cortes C, Vapnik V (1995) Support-vector networks. Machine learning 20(3):273–297. doi:10. 1023/A:1022627411411

16. Coutaz J, Nigay L, Salber D, Blandford A, May J, Young RM (1995) Four easy pieces for assessing the usability of multimodal interaction: the CARE properties. In: Proceedings of INTERACT95, Lillehammer, pp 115–120

17. Cowie R, Douglas-Cowie E, Tsapatsoulis N, Votsis G, Kollias S, Fellenz W, Taylor J (2001) Emotion recognition in human-computer interaction. IEEE Signal Process Mag 18(1):32–80. doi:10.1109/79.911197

18. Darwin C (1872) The expression of the emotions in man and animals, 1st edn. Oxford University Press Inc, New York

19. Dempster AP (1968) A generalization of bayesian inference. J R Stat Soc Ser B (Methodological), pp 205–247

20. Devillers L, Vidrascu L, Lamel L (2005) Challenges in real-life emotion annotation and machine learning based detection. Neural Netw 18(4):407–422. doi:10.1016/j.neunet.2005. 03.007 (emotion and Brain)

21. Dey AK, Abowd GD (1999) Towards a better understanding of context and context-awareness. In: HUC '99: Proceedings of the 1st international symposium on handheld and ubiquitous computing, Springer, Berlin, pp 304–307

22. Dietrich C, Schwenker F, Palm G (2001) Classification of time series utilizing temporal and decision fusion. In: Multiple classifier systems, Springer, Berlin, pp 378–387

23. Ekman P (1992) An argument for basic emotions. Cognit Emot 6(3–4):169–200. doi:10.1080/ 02699939208411068

24. Ekman P, Friesen WV (1978) Facial action coding system: a technique for the measurement of facial movement. Consulting Psychologists Press, Palo Alto

25. Endrass B, Haering M, Akila G, André E (2014) Simulating deceptive cues of joy in humanoid robots. In: Bickmore T, Marsella S, Sidner C (eds) Intelligent virtual agents, Lecture notes in computer science, vol 8637, Springer, Berlin, pp 174–177. doi:10.1007/978-3-319-09767-1_ 20

26. Frijda NH (1994) Varieties of affect: emotions and episodes, moods, and sentiments. In: Ekman P, Davidson RJ (eds) The nature of emotion, fundamental questions. Oxford University Press, New York, pp 197–202

27. de Gelder B (2006) Towards the neurobiology of emotional body language. Nat Rev Neurosci 7(3):242–249. doi:10.1038/nrn1872

28. Glodek M, Scherer S, Schwenker F, Palm G (2011) Conditioned hidden markov model fusion for multimodal classification. In: INTERSPEECH, pp 2269–2272

29. Glodek M, Schels M, Schwenker F, Palm G (2014) Combination of sequential class distributions from multiple channels using markov fusion networks. J Multimodal User Interfaces 8(3):257– 272

30. Glodek M, Honold F, Geier T, Krell G, Nothdurft F, Reuter S, Schüssel F, Hörnle T, Dietmayer K, Minker W, Biundo S, Weber M, Palm G, Schwenker F (2015) Fusion paradigms in cognitive technical systems for human-computer interaction. Neurocomputing 161:17–37. doi:10.1016/ j.neucom.2015.01.076

31. Gram C, Cockton G (1997) Design principles for interactive software. Chapman & Hall Ltd, London

32. Guyon I, Elisseeff A (2003) An introduction to variable and feature selection. J Mach Learn Res 3:1157–1182
33. Honold F, Schüssel F, Panayotova K, Weber M (2012) The nonverbal toolkit: Towards a framework for automatic integration of nonverbal communication into virtual environments. In: 2012 8th international conference on intelligent environments (IE), pp 243–250. doi:10.1109/IE.2012.13
34. Honold F, Schüssel F, Weber M (2012) Adaptive probabilistic fission for multimodal systems. In: Proceedings of the 24th Australian computer-human interaction conference, OzCHI '12, ACM, New York, pp 222–231. doi:10.1145/2414536.2414575
35. Honold F, Schüssel F, Weber M, Nothdurft F, Bertrand G, Minker W (2013) Context models for adaptive dialogs and multimodal interaction. In: 2013 9th International conference on intelligent environments (IE), IEEE, pp 57–64. doi:10.1109/IE.2013.54
36. Honold F, Bercher P, Richter F, Nothdurft F, Geier T, Barth R, Hörnle T, Schüssel F, Reuter S, Rau M, Bertrand G, Seegebarth B, Kurzok P, Schattenberg B, Minker W, Weber M, Biundo S (2014) Companion-technology: towards user—and situation-adaptive functionality of technical systems. In: 2014 10th International conference on, intelligent environments (IE), IEEE, pp 378–381. doi:10.1109/IE.2014.60
37. Honold F, Schüssel F, Weber M (2014) The automated interplay of multimodal fission and fusion in adaptive HCI. In: 2014 10th International conference on intelligent environments (IE), IEEE, China, pp 170–177. doi:10.1109/IE.2014.32
38. Itakura F (1975) Line spectrum representation of linear predictor coefficients of speech signals. J Acoust Soc Am 57(S1):S35–S35. doi:10.1121/1.1995189
39. Joachims T (2006) Transductive support vector machines. Chapelle et al. (2006), pp 105–118
40. Jolliffe I (2002) Principal component analysis. Wiley, New York. Online Library
41. Kächele M, Zharkov D, Meudt S, Schwenker F (2014) Prosodic, spectral and voice quality feature selection using a long-term stopping criterion for audio-based emotion recognition. In: 22nd international conference on pattern recognition, ICPR 2014, Stockholm, pp 803–808, 24–28 August 2014. doi:10.1109/ICPR.2014.148
42. Kandel E, Schwartz J, Jessell T (2000) Principles of neural science. McGraw-Hill, New York
43. Kelley JF (1984) An iterative design methodology for user-friendly natural language office information applications. ACM Trans Inf Syst 2(1):26–41
44. Lisetti CL, Schiano DJ (2000) Automatic facial expression interpretation: where human-computer interaction, artificial intelligence and cognitive science intersect. Pragmat Cognit (Spec Issue Fac Inf Process Multidiscip Perspect) 8(1):185–235
45. Littlewort G, Whitehill J, Wu T, Fasel I, Frank M, Movellan J, Bartlett M (2011) The computer expression recognition toolbox (cert). In: 2011 IEEE international conference on automatic face gesture recognition and workshops (FG 2011), pp 298–305. doi:10.1109/FG.2011.5771414
46. Logan B (2000) Mel frequency cepstral coefficients for music modeling. In: ISMIR
47. Lugger M, Yang B, Wokurek W (2006) Robust estimation of voice quality parameters under realworld disturbances. In: 2006 IEEE international conference on acoustics, speech and signal processing, 2006. ICASSP 2006 Proceedings, vol 1, pp 1097–1100
48. Matsugu M, Mori K, Mitari Y, Kaneda Y (2003) Subject independent facial expression recognition with robust face detection using a convolutional neural network. Neural Netw 16(5):555–559
49. Meudt S, Schwenker F (2012) On instance selection in audio based emotion recognition. In: Proceedings, of the artificial neural networks in pattern recognition—5th INNS IAPR TC 3 GIRPR Workshop, ANNPR 2012, Trento, pp 186–192, 17–19 September 2012. doi:10.1007/978-3-642-33212-8_17
50. Meudt S, Bigalke L, Schwenker F (2012) Atlas—an annotation tool for hci data utilizing machine learning methods. Proceedings of the 1st international conference on affective and pleasurable design (APD'12) [jointly with the 4th international conference on applied human factors and ergonomics (AHFE'12)]. CRC Press, Advances in human factors and ergonomics series, pp 5347–5352

51. Mitra S, Acharya T (2007) Gesture recognition: a survey. IEEE Trans Syst Man Cybern Part C: Appl Rev 37(3):311–324. doi:10.1109/TSMCC.2007.893280
52. Nothdurft F, Bertrand G, Heinroth T, Minker W (2010) Geedi—guards for emotional and explanatory dialogues. In: Callaghan V, Kameas A, Egerton S, Satoh I, Weber M (eds) 2010 Sixth international conference on intelligent environments, IEEE Computer Society, pp 90–95. doi:10.1109/IE.2010.24
53. Ojala T, Pietikäinen M, Harwood D (1996) A comparative study of texture measures with classification based on featured distributions. Patt Recognit 29(1):51–59. doi:10.1016/0031-3203(95)00067-4
54. Oudeyer PY (2003) The production and recognition of emotions in speech: features and algorithms. Int J Hum-Comput Stud 59(1–2):157–183. doi:10.1016/S1071-5819(02)00141-6
55. Picard RW (1995) Affective Computing. M.I.T media laboratory perceptual computing section, Technical Report No 321
56. Picard RW, Vyzas E, Healey J (2001) Toward machine emotional intelligence: analysis of affective physiological state. IEEE Trans Patt Anal Mach Intell 23(10):1175–1191. doi:10. 1109/34.954607
57. Plutchik R, Kellerman H (1980) Theories of emotion. Emotion: theory, research, and experience, Academic press INC, USA
58. Rabiner L (1989) A tutorial on hidden markov models and selected applications in speech recognition. Proceedings of the IEEE 77(2):257–286. doi:10.1109/5.18626
59. Rabiner L, Juang BH (1993) Fundamentals of speech recognition. Prentice Hall PTR, New Jersey
60. Raybourn EM (2014) A new paradigm for serious games: transmedia learning for more effective training and education. J Comput Sci 5(3):471–481
61. Russell JA, Mehrabian A (1977) Evidence for a three-factor theory of emotions. J Res Pers 11(3):273–294
62. Schaub FM (2014) Dynamic privacy adaptation in ubiquitous computing. Dissertation, Universität Ulm. Fakultät für Ingenieurwissenschaften und Informatik. http://vts.uni-ulm.de/doc. asp?id=9029
63. Scherer K (2005) What are emotions? and how can they be measured? Soc Sci Inf 44(4):695–729. doi:10.1177/0539018405058216
64. Scherer K, Shuman V, Fontaine J, Soriano C (2013) The grid meets the wheel: assessing emotional feeling via self-report. In: Fontaine J, Scherer K, Soriano C (eds) Components of emotional meaning : a sourcebook. Oxford University Press, Series in affective science, pp 281–298
65. Scherer S, Schwenker F, Palm G (2009) Classifier fusion for emotion recognition from speech. In: Advanced intelligent environments, Springer, Berlin, pp 95–117
66. Scherer S, Kane J, Gobl C, Schwenker F (2013) Investigating fuzzy-input fuzzy-output support vector machines for robust voice quality classification. Comput Speech Lang 27(1):263–287, doi:10.1016/j.csl.2012.06.001 (Special issue on paralinguistics in naturalistic speech and language)
67. Schmidt M, Schels M, Schwenker F (2010) A hidden markov model based approach for facial expression recognition in image sequences. In: Proceedings of the 4th IAPR TC3 workshop on artificial neural networks in pattern recognition (ANNPR'10), LNAI 5998, pp 149–160. doi:10.1007/978-3-642-12159-3_14
68. Schüssel F, Honold F, Weber M, Schmidt M, Bubalo N, Huckauf A (2014) Multimodal interaction history and its use in error detection and recovery. In: Proceedings of the 16th ACM international conference on multimodal interaction, ICMI '14, ACM, New York, pp 164–171. doi:10.1145/2663204.2663255
69. Schwenker F, Scherer S, Schmidt M, Schels M, Glodek M (2010) Multiple classifier systems for the recogonition of human emotions. In: El Gayar N, Kittler J, Roli F (eds) Multiple classifier systems, Lecture notes in computer science, vol 5997, Springer, Berlin, pp 315–324. doi:10. 1007/978-3-642-12127-2_33

70. Shafer G, et al. (1976) A mathematical theory of evidence, vol 1. Princeton University Press, USA
71. Shneiderman B (2007) Foreword. In: Sears A, Jacko JA (eds) The human-computer interaction handbook: fundamentals, evolving technologies and emerging applications, Second edition (Human factors and ergonomics), 2nd edn, CRC Press, pp XIX-XX
72. Stuhlsatz A, Meyer C, Eyben F, Zielke T, Meier G, Schuller B (2011) Deep neural networks for acoustic emotion recognition: raising the benchmarks. In: 2011 IEEE International Conference on Acoustics, speech and signal processing (ICASSP), IEEE, pp 5688–5691
73. Sun D, Roth S, Black M (2010) Secrets of optical flow estimation and their principles. In: 2010 IEEE conference on Computer vision and pattern recognition (CVPR), pp 2432–2439. doi:10.1109/CVPR.2010.5539939
74. Thiel C, Scherer S, Schwenker F (2007) Fuzzy-input fuzzy-output one-against-all support vector machines. In: Knowledge-based intelligent information and engineering systems, Springer, Berlin, pp 156–165
75. Valstar M, Schuller B, Smith K, Eyben F, Jiang B, Bilakhia S, Schnieder S, Cowie R, Pantic M (2013) Avec 2013: The continuous audio/visual emotion and depression recognition challenge. In: Proceedings of the 3rd ACM international workshop on audio/visual emotion challenge, AVEC '13, ACM, New York, pp 3–10. doi:10.1145/2512530.2512533
76. Vogt T, André E (2005) Comparing feature sets for acted and spontaneous speech in view of automatic emotion recognition. In: IEEE international conference on multimedia and expo, 2005. ICME 2005, pp 474–477. doi:10.1109/ICME.2005.1521463
77. Walter S, Scherer S, Schels M, Glodek M, Hrabal D, Schmidt M, Böck R, Limbrecht K, Traue HC, Schwenker F (2011) Multimodal emotion classification in naturalistic user behavior. In: Jacko JA (ed) Proceedings of the 14th international conference on human computer interaction (HCI'11), Springer, LNCS 6763, pp 603–611
78. Walter S, Kim J, Hrabal D, Crawcour S, Kessler H, Traue H (2013) Transsituational individual-specific biopsychological classification of emotions. IEEE Trans Syst Man Cybern 43(4):988–995
79. Weiser M (1999) The computer for the 21st century. SIGMOBILE Mob Comput Commun Rev (This article first appeared in Scientific America) 3(3):3–11, 1991. doi:10.1145/329124. 329126, vol 265, no 3 (September 1991), pp 94–104
80. Wendemuth A, Biundo S (2012) A companion technology for cognitive technical systems. In: Esposito A, Esposito AM, Vinciarelli A, Hoffmann R, Müller VC (eds) Cognitive behavioural systems, LNCS, vol 7403, Springer, Berlin, pp 89–103. doi:10.1007/978-3-642-34584-5_7
81. Wundt W (1896) Grundriss der psychologie. Engelmann, Leipzig
82. Yang J, Yan R, Hauptmann AG (2007) Cross-domain video concept detection using adaptive svms. In: Proceedings of the 15th international conference on multimedia, ACM, pp 188–197
83. Zhao G, Pietikainen M (2007) Dynamic texture recognition using local binary patterns with an application to facial expressions. IEEE Trans Patt Anal Mach Intell, 29(6):915–928. doi:10.1109/TPAMI.2007.1110

Chapter 7
Physical and Moral Disgust in Socially Believable Behaving Systems in Different Cultures

Barbara Lewandowska-Tomaszczyk and Paul A. Wilson

Abstract The aim of the present study is to use the GRID, online emotions sorting and corpus methodologies to illuminate different types of disgust that an emotion-sensitive socially interacting robot would need to encode and decode in order to competently produce and recognise these and other types of physical, moral and aesthetic types of complex emotions in social settings. We argue that emotions in general, and different types of disgust as an instance of these, differ with respect to the amount of *cognitive grounding* they need in order to arise and social robots will successfully use such emotions provided they do not only recognise and produce physical, bodily manifestations of emotions, but also have access to large knowledge bases and are able to process situational context clues. The different types of disgust are identified and compared cross-culturally to provide an evaluation of their relative salience. The study also underscores the conceptual viewpoint of emotions as clusters of emotions rather than solitary, individual representations. We argue that such clustering should be at the heart of emotions modelling in social robots. In order to successfully use the emotion of disgust in their interactions with humans, robots need to be sensitive to possible within-culture and cross-culture differences pertaining to such emotions, exemplified by British English and Polish in the present study. Given the centrality of values to the emotion of disgust, robots need to have the capacity to update from a knowledge base and learn from the situational context the set of values for each significant human that they interact with.

Keywords Aesthetics · British English · *Disgust* cluster · Emotion event (scenario) · GRID · Language corpora · Moral emotions · Online emotions sorting study · Polish · Social robots · Translation data · *Wstręt* cluster

B. Lewandowska-Tomaszczyk (✉)
State University of Applied Sciences in Konin, University of Lodz, Lodz, Poland
e-mail: blt@uni.lodz.pl

P.A. Wilson
University of Lodz, Lodz, Poland

© Springer International Publishing Switzerland 2016
A. Esposito and L.C. Jain (eds.), *Toward Robotic Socially Believable Behaving Systems - Volume I*, Intelligent Systems Reference Library 105,
DOI 10.1007/978-3-319-31056-5_7

7.1   Introduction

7.1.1   Role of Emotions in Social Robotics

The success that robots achieve in their assistance of human functioning in everyday life depends on the extent to which they are able to interact with humans at a social level. The arguments aptly presented in Hudlicka and McNeese [13] explicitly show the reasons why and to what extent artificial agents (robots) should develop the properties which are judged as intentional and expressive when interacting with humans. It is socially believable robots, expressing emotions and perceiving emotional states in humans, which will be perceived as acceptable social partners in interaction.

Emotions also serve regulatory mechanisms in social robots and influence their adaptive function. One of the most crucial challenges facing the adaptive functions of emotion–sensitive interactive robots is their ability to interact socially in different languages and cultures. Our recent observation of differences in the profiles of joy, sadness, fear and anger in British English and Polish show that culture-specific modelling of affective behaviour is a crucial property of socially believable behaving systems [34]. Being at the centre of morals, values and rights violations, the present study into British English versus Polish representations of disgust places social robotics at the heart of issues that are of fundamental importance to human life.

7.1.2   Typology of Disgust

7.1.2.1   Disgust Event Scenarios

In Lewandowska-Tomaszczyk and Wilson [19] we propose that emotions arise in the context of particular Emotion Events as in the following structure of a prototypical Emotion Event scenario:

> **Prototypical Emotion Event Scenario EES:** Context (*Biological* predispositions of Experiencer, *Social* and *Cultural* conditioning, *On-line* contextual properties of Event) [Stimulus → Experiencer{(internally experienced) Emotion [**EMBODIED mind—METAPHOR**] + (internally and externally manifested) physiological and physical symptoms} → possible external reaction(s) of Experiencer (approach/avoidance, language) **i.e., EXBODIED mind**

Extended Emotion Event Scenarios involve cases of experiencing more than one emotion of *the same valence* at the same time, i.e., *emotion clusters* on the one hand, and so-called *mixed feelings*, experienced as two *contradictory emotions* at the same time on the other.

Disgust, etymologically derived from Latin in the original meaning 'exposed to/in the way of harm, injury, danger', arises in a conceptual space inhabited by a cluster of related emotions such as *aversion, revulsion, repulsion* (Pol. *awersja, zniesmaczenie, obrzydzenie, wstręt, niesmak*). They are situated either within the same cluster or connected to other emotions and emotion clusters of the same negative polarity such as anxiety, terror, anger, pain, gloom, displeasure, self-loathing, contempt and fear. Such chains of senses, either *intra-* or *inter-cluster members*, have a significant frequency of occurrence in the analysed corpus data.

Disgust is characterised by a combination of facial gestures, body posture and movements as well as behavioural features, some of them difficult to realistically reproduce with robots but less problematic for recognition. When such features appear with the human agent they will serve as plausible sources of information for the interacting robot to identify a particular emotion or rather emotion cluster.

We assume that in many cases both the expression and recognition of a particular emotion is involved in a whole cluster of emotions related strongly enough to be perceived as one category. The materials we consider in our research are both bodily reactions as identified in questionnaires as well as large collections of language data (corpora), which serve as a source to acquire vocabulary involving both emotional language, i.e., language used in an emotional state, as well as language referring to and describing individual emotions and their conditioning in terms of an Emotion Event scenario [19].

7.1.2.2   Physical Versus Moral Disgust

Our claim argued for in the present paper refers to a distinction between more phys-ically based and morality based emotions on the example of the emotion of disgust. One of the ways in which physical disgust differs from moral disgust is in terms of cognitive processing. For example, the model advanced by Ortony et al. [23] argues that disgust (and surprise) do not require much cognitive processing. We propose that while this applies to physical disgust, moral disgust requires background infor-mation and knowledge of the outside world. Moral emotions such as disgust then require wider and deeper cognitive grounding, expressed in terms of knowledge of a particular culture and the experiencer's beliefs and preferences than, for example, fear.[1]

The extent to which physical and moral disgust can be regarded as valid instances of disgust has been questioned. One viewpoint is that it is only physical disgust that is a true emotion and that moral disgust is a metaphor of this that is used to convey indignation, resentment or disapproval [14, 26]. By referring to moral disgust as *indignation*, Moll et al. [21] highlight that it is conceptually related to but distinct from disgust.

Other theoretical stances, as Lee and Ellsworth [18] explain, propose that physi-cal and moral disgust are two valid but distinct clusters of disgust that have distinct conceptual proximities to anger and fear. Using the GRID methodology they showed

---

[1]However, as argued in Lewandowska-Tomaszczyk and Wilson [19] and Wilson and Lewandowska-Tomaszczyk [34], even such basic emotions as fear, sadness, joy or anger cannot be considered fully universal emotions, devoid of cultural impact.

that whereas moral disgust is characterised relatively more by anger, there is a relatively greater salience of fear in physical disgust, which evolved to enhance the protection of the organism from germs, parasites and contaminated food [28]. More specifically, it was found that moral disgust and anger are characterised by social norm violation and features pertaining to an orientation towards others and punishment. In contrast, physical disgust and fear are more associated with weakness, submissiveness, avoidance and compliance. In a more refined examination of moral disgust in a study in which participants' responses to verbal descriptions of a norm violation involving either the body (bodily moral) or rights and harm (socio-moral), Gutierrez, Giner-Sorolla and Vasiljevic [10] observed "that anger was greater when socio-moral harm was described, and disgust was greater when bodily moral violations were described" (p. 60). *Bodily moral disgust* concerns disgust that occurs in response to norm violation associated with the body and includes issues such as incest and homosexuality and other departures from what is seen as sexual norms, as well as societal standards regarding diet, including the types and parts of animals that are considered part of the dietary norm. The violation of rights and moral norms is characterised by *socio-moral disgust*, which pertains to values such as equality, trust, exploitation, dishonesty, betrayal, racism and sadism. All in all, whereas physical and moral disgust are characterised relatively more by fear and anger, respectively, a more precise focus on moral disgust reveals the possibility that whereas *bodily moral disgust* is associated more with disgust, *socio-moral disgust* is conceptually closer to anger.

7.1.2.3 Examples of Disgust Types

To explain the distinction more fully, whereas physical disgust involves responses to physical objects and situations, either as biological, almost instinctive, reactions to their physical properties, moral disgust occurs in response to infringements of rights and ethical or moral laws and norms. Within what we might refer to as 'moral scenarios' are those which are clearly infringements of social rules (e.g., the maltreatment of animals), although not necessarily moral rules in the prototypical sense. Morality is an extremely fuzzy notion—for example, in the Pussy Riot, nakedness in public places—particularly churches—was certainly provocative and breaking the religious community's code of conduct. In a wider religious sense, as proposed by Brinkmann and Musaeus [3], "disgust is related to the code of divinity: When what is pure and sacred is polluted and degraded, disgust is the relevant emotional reaction. Clear examples are found in relation to the (cultural and religious) normativities around food, but also many other degradations of the sacred can be found" (p. 135). Other non-religious examples can include milder cases—our students not coming to classes because of their laziness or a child being late to school or not brushing his teeth. We accept such cases as instances of immorality, which shows how context specific and dynamic this concept is. It also means that we assume morality to be synonymous with any type of social rules of a general and community-bound kind.

In this respect, aesthetic disgust can be considered to be related to moral disgust, especially with reference to abstract objects.

A typology of disgust can thus be proposed accordingly, exemplified by authentic data:

(Physical) disgust, which appears with reference to physical objects as stimuli.

(1) *he had the most **disgusting** rotten teeth*
(2) (culture-specific) *if people eat food which is thought of as **disgusting** and unclean what are these people like?*
(3) *After a few weeks on an unsweetened diet, it is remarkable how **disgusting** anything sugary tastes*

(Ethical/Moral) disgust, which typically appears with reference to abstract concepts and events. Physical objects (animates, things) can function as stimuli of moral disgust by metonymy (i.e., a figure of thought in which they conceptually stand for a whole event or an abstract concept).

(4) *down the gangplank complaining vaguely to his friends that there had been **disgusting** favouritism somewhere*
(5) *I do have slight vegetarian principles like I think it's **disgusting** to use animals for luxury things like cosmetics and fur coats*
(6) *which had earlier tended to be pro-Nazi in their sympathies—stated that a **disgust** about the Part was building up among the people*

(Aesthetic) disgust, which is an exbodiment of the experiencer's judgment concerning beauty versus ugliness.

(7) *We eat with our fingers—they think that is dirty and **disgusting***
(8) *pork pie and cold sausages, occasionally varied by Albert's own cooking in a **disgusting** frying pan when the spirit moved him*

Aesthetic disgust is related to moral disgust when it refers to abstract objects.

(9) *I think you're like some **disgusting** little animal, some creature from another planet*
(10) *At 40, you developed something flabby, **disgusting** and unavoidable called middle-aged spread and your waist disappearing along ...*
(11) *They would probably play that **disgusting** game of spitting on people from a great height*

Mixed types of disgust (physical, ethical, aesthetic).

(12) *The prevailing mood is **disgust**; at the pathetic inarticulacy of the great ape on a rugby field, Frank Machin ...*
(13) *ritual slaughter scenes—some fainting, and others leaving the cinema in **disgust***

7.1.2.4 Metaphor and Disgust

Disgust is one of the basic emotion categories. In its physical sense it is also evidenced in the majority of languages as the most characteristic metaphoric pattern for moral disgust DISGUST IS NAUSEA [16]:

Alternate names: MENTAL DISAFFECTION IS UNPLEASANT VISCERAL REACTION
Source Domain: nausea
Target Domain: disgust

Examples:
You make me sick!
His behaviour made me want to throw up.
She's a nauseating person.
It's obnoxious (*obnoxious* is etymologically related to harmful, poisonous).

Therefore, mentally disgusting or repulsive stimuli bring about similar biological and cognitive reactions to those present with regards to physically disgusting food or other objects considered obnoxious (political issues, social judgments), even those which would be considered acceptable and even desired by others.[2]

7.1.3 Overview of Present Study

In the present paper we investigate disgust scenarios in British English and Polish as instantiations of two languages and cultures. Our project makes an attempt to propose a discussion of a model in which emotions have a potential role in *action selection* with artificial systems. Emotions function as monitoring and regulatory mechanisms by means of behaviour modelling parameters and cognitive-evaluative system (morality)—the robot performs behaviour selection on the basis of these two types of mechanisms to retain stability (homeostasis) and viability (survival). The human or artificial agent, when experiencing an emotion, shows either a behavioural reaction in the sense of pursuing a goal (approach reaction), or withdrawing (avoidance strategy), or displaying a still more complex behaviour [19]. Disgust is one of the emotions which results in the avoidance, withdrawal, rejection strategy. Contrary to Cañamero and Gaussier [4], who do not identify disgust as one of the six crucial mechanisms for behaviour adaptation, it is argued here that it is disgust which combines two facets of the monitoring mechanisms—physical and moral (ethical and aesthetic) and it is precisely disgust as one of the basic mechanisms that helps to monitor the selection as opposed to the rejection of the ensuing action as a result (e.g., human aversion to endorse harm in simulating murder in Cushman et al. [5]).

---

[2]A related viewpoint is that of a semantic framework of impoliteness incorporated into a system of communicative behaviour management [32], which builds on the argument that (im)politeness behaviours are manifestations of offence management, where offence is rooted in disgust.

A fundamental element of the present study is the identification of core features of disgust as produced and identified cross-culturally on the example of British English and Polish. The other challenge is the problem of the identification of core criterial properties of individual emotions. We propose in Lewandowska-Tomaszczyk and Wilson [19] that both the production and recognition of emotions are products of emotion clustering rather than single emotions. In a study by Breazeal [2], who administered a task that required participants to choose, from a number of basic emotions, those emotions that provided the best match of facial expressions of pictured robots, the recognition rate with respect to disgust was 70.6 %, while 17.6 % of subjects recognised disgust as anger, 5.9 % as fear, and 5.9 % as sternness (unfriendliness, displeasure). However, as emotion expression is conveyed through the entire body as well as emotional language, what we aim to present in the subsequent sections is a comparison between behavioural, cognitive and linguistic properties in two different cultures, British English and Polish, to achieve less fuzzy profiles of the emotion in cross-cultural settings. As will be argued in the present paper, the properties of the basic emotion of disgust in British English are only asymmetrically related to its putative lexicographic equivalents in Polish. Modelling of socially believable behaviour of robots in the affective domain needs to take these differences into consideration.

7.1.3.1 Aims

The present study underscores the conceptual viewpoint of emotions as clusters of emotions rather than distinct, unified representations. We argue that such clustering should be at the heart of emotions modelling in social robots. It is also clear that if such robots are to successfully use the emotion of disgust in their interactions with humans they need to be sensitive to possible within-culture and cross-culture differences pertaining to this emotion. Given the centrality of values to the emotion of disgust, they would also need to mimic humans in their capacity to, for example, update and learn the set of values for each significant human that they interact with.

The main aim is to employ the GRID, online emotions sorting and corpus methodologies to compare the different types of disgust that an emotion–sensitive socially interacting robot would need to encode and decode in order to competently use such variants in social settings in Britain vis-à-vis Poland. The different types of disgust (*disgust* and *revulsion* in British English, and *wstręt* 'repulsion, revulsion', *obrzydzenie* 'repulsion, disgust, revulsion' and *zniesmaczenie* 'disgust' in Polish) will be compared to provide an evaluation of their relative salience and importance in the everyday lives of human beings. A major focus of this comparison will be on the relative salience of physical versus moral disgust in each of the disgust types, which will be determined by withdrawal orientation and moral transgression as well as, following Lee and Ellsworth [18], their conceptual proximity to fear and anger, respectively.

7.2 Materials and Methods

The comparison of disgust scenarios in British English and Polish was achieved with the use of three complementary methodological paradigms: the online emotions sorting, GRID, and cognitive corpus linguistics methodologies.

7.2.1 Online Emotions Sorting Methodology

In the emotions sorting methodology, emotion terms are typically presented simultaneously on a desk in front of participants who are free to categorise them into as many or as few groups as they wish. In the online version the sorting takes place on the computer desktop. In the only study employing the category sorting task in a cross-cultural perspective (to our knowledge), the conceptual structure of Dutch versus Indonesian was investigated by Fontaine et al. [8]. The category sorting task has also been used to determine the conceptual structure of a specific emotion feature—pleasure [6].

In the present study we adapted the NodeXL [29] tool to provide information pertaining to the disgust cluster in British English and Polish and the relationship between each of these to their respective anger and fear clusters. Although the most common use of NodeXL is to analyse relationships between individuals using online social media networks, we employ NodeXL to create graphical representations of the Polish and British English co-occurrence emotion matrices. The graphs created are similar to those produced by the synonyms rating methodology employed by Heider [12] to compare and contrast emotion terms across three Indonesian languages. In Heider's [12] study participants provided a synonym emotion for each target emotion term and in the maps of the emotion domains the nodes are represented by the individual emotion terms. For the sake of consistency, we adopt the same terminology as Heider [12] where possible. The main difference is in the terms used to refer to the links that show the relationships between the nodes; whereas for Heider connection strength refers to the between-subjects frequency with which an emotion term is given as the synonym for another, the connection strength in our NodeXL graphs represent the co-occurring frequency of the emotion terms in the online emotions sorting data, and are hence sometimes referred to as *co-occurrence values* or *interconnections*.

7.2.1.1 Procedure

Participants volunteered to take part in the study either through direct contact by one of the authors or in response to adverts placed on internet forums. Each volunteer was sent a link to the experimental platform and was allowed to take part in the experiment at a time and location of their choosing, with the request that they do the experiment in seclusion. The first page presented the British and Polish flags and the participants clicked on these according to their nationality. Then the instructions page appeared in the appropriate language. Initially, there was a brief introduction outlining that the study was concerned with finding out about how people think some emotions "go together" and other emotions belong in different categories. More detailed instructions regarding the specific sorting task were as follows:

> You will be presented with 135 emotions on the computer screen. We'd like you to sort these emotions into categories representing your best judgement about which emotions are similar to each other and which are different from each other. There is no one correct way to sort the emotions - make as few or as many categories as you wish and put as few or as many emotions in each group as you see fit. This study requires careful thought and you therefore need to carefully think about which category each emotion belongs rather than just quickly putting emotions in categories without much thought.

Following this, participants were told they would watch a video (about 8 min) that would demonstrate the procedure. They were told that this would be followed by a practice session that involved the categorisation of food items, and once this had been completed the proper experiment with emotion terms would begin. The following message appeared in a central window on the experimental page:

> You need to click on the "New Emotions Group" button and drag emotions to create your emotion groups. When you have finished creating your emotion groups, click on the orange "DONE" button and the experiment has been completed.

7.2.1.2   Participants

There were 58 British English participants (27 females, mean age = 42.7 years) with the following occupations: academic departmental manager, administrator (3), civil servant, cleaner, company director, IT (4), consultant (3), editor, events manager, executive coach, housewife (3), lecturer (5), manager, psychologist (2), radiographer, retired (6), tailoress, scientist, self-employed, student (11), supported housing officer, teacher (4), teaching assistant, unemployed, volunteer (one participant did not state their occupation). There were 58 Polish participants (27 females, mean age = 35.8 years) with the following occupations: account manager, accountant, career advisor, cashier, cultural studies specialist, doctor, IT (3), lecturer (14), marketing employee, office employee (2), pedagogist, project manager, psychologist, student (10), teacher (16), translator (3).

7.2.2   GRID

The GRID instrument [9, 27] employs a system of dimensions and components, which bring about insight into the nature of emotion prototypical structures. The GRID project is coordinated by the Swiss Center for Affective Sciences at the University of Geneva in collaboration with Ghent University and is a worldwide study of emotional patterning across 23 languages and 27 countries. The GRID instrument comprises a Web-based questionnaire in which 24 prototypical emotion terms are evaluated on 144 emotion features. These features represent activity in all six of the major components of emotion. Thirty-one features relate to appraisals of events, eighteen to psychophysiological changes, twenty-six to facial, vocal or gestural expressions, forty to action tendencies, twenty-two to subjective experiences and four to emotion regulation. An additional three features refer to other qualities, such as frequency and social acceptability of the emotion. Participants are asked to rate the likelihood of these features for the various emotions. This methodology is comprehensive in its scope as it allows the multicultural comparison of emotion conceptualisations on all six of the emotion categories recognised by emotion theorists [7, 22, 27].

7.2.2.1 Procedure

Participants completed the GRID instrument in a controlled Web study [25], in which each participant was presented with four emotion terms randomly chosen from the set of 24 and asked to rate each in terms of the 144 emotion features. They rated the likelihood that each of the 144 emotion features can be inferred when a person from their cultural group uses the emotion term to describe an emotional experience. A 9-point scale was employed that ranged from extremely unlikely (1) to extremely likely (9)—the numbers 2–8 were placed at equidistant intervals between the two ends of the scale, with 5 'neither unlikely, nor likely' in the middle and participants typed their ratings on the keyboard. It was clearly stated that the participants needed to rate the likelihood of occurrence of each of the features when somebody who speaks their language describes an emotional experience with the emotion terms presented. Each of the 144 emotion features was presented separately, and participants rated all four emotion terms for that feature before proceeding to the next feature.

7.2.2.2 Participants

The mean ages and gender ratios of the participants for each of the emotion terms were as follows: *disgust* (35 British English-speaking participants; mean age 21.9 years, 18 females); *wstręt* (25 Polish-speaking participants; mean age 21.8 years, 18 females).

7.2.3 Language Corpus Data

In order to extract the context of the use of emotion terms in English and Polish, we resort to large language corpus data, particularly collocations (words co-occurring more frequently than by chance, minimally 5 times in the materials consulted) and their frequencies. Two selected *association scores* are computed for each collocational combination: *t*-score (TTEST) and mutual information (MI). By analysing authentic language we can detect shifts in meaning for the same linguistic form and we can also describe the contexts which support such shifts. Based on the frequency of occurrence, corpus-based methods let us statistically determine which linguistic meanings are most salient. The materials we use come from the British National Corpus (100 million words) and the National Corpus of Polish (nkjp.pl), which contains 300 million units of balanced data, to which we apply the process of (corpus size) normalisation to enable comparison across these differently sized large datasets. Polish-to-English and English-to-Polish translation corpora and the tools of their alignment (http://clarin.pelcra.pl/Paralela/) show a number of cross-linguistic correspondences within the disgust cluster. The materials contained in the BNC and NKJP are structurally comparable to a large extent and contain language from similar domains, styles and genres, although the narrow contexts in which the forms of the *disgust*/*wstręt* clusters are used are not always identical in the two languages. The quantitative data are analysed using cognitive linguistic instruments, particularly with reference to *construal* [17] of *disgust* event scenarios as expressed by structural properties of the language involved and *metaphors* and other figurative language uses [15].

7.3 Results

7.3.1 Online Emotions Sorting Results

7.3.1.1 Intra-linguistic Differences

In terms of Lee and Ellsworth's [18] observations of greater conceptual contiguity between physical disgust and fear and between moral disgust and anger the interconnections between disgust in Polish and British English and fear and anger suggest a greater salience of moral disgust in both languages. This is shown in the relatively higher co-occurrences between the anger cluster emotions and both the British English disgust emotions (e.g., *anger—disgust* (32) and *anger—revulsion* (25)) and the Polish disgust emotions (e.g., *złość* 'anger 1'–*wstręt* 'repulsion, revulsion' (25), *złość* 'anger 1'—*obrzydzenie* 'repulsion, disgust, revulsion' (18) and *złość* 'anger 1'—*zniesmaczenie* 'disgust' (8)). In contrast, the relatively lower interconnections between the fear cluster emotions and both the British English disgust emotions (e.g., *fear—disgust* (9) and *fear—revulsion* (6)) and the Polish disgust emotions (e.g., *strach* 'fear'—*wstręt* 'repulsion, revulsion' (14), *strach* 'fear'—*obrzydzenie* 'repulsion, disgust, revulsion' (11) and *strach* 'fear'—*zniesmaczenie* 'disgust' (4)) show that disgust is characterised by relatively less physical disgust in both languages (see Figs. 7.1, 7.2 and 7.3). The only caveat to be highlighted is that *zniesmaczenie* 'disgust' has more of a relative balance in terms of moral and physical disgust than the other types of disgust.

Fig. 7.1 Co-occurrence values between British English disgust and anger and fear clusters

7.3.1.2 British English Versus Polish Disgust—Anger Cluster Co-Occurrences

It can be seen in Fig. 7.1 that *disgust* has relatively high co-occurrences with British English anger cluster emotions (e.g., *wrath* (34), *fury* (34) and *anger* (32)). In contrast, there are lower interconnections between *wstręt* 'repulsion, revulsion' and Polish anger cluster emotions (e.g., *furia* 'fury' (23), *złość* 'anger 1' (25) and *gniew* 'anger 2' (20)) and between *obrzydzenie* 'repulsion, disgust, revulsion' and Polish anger cluster emotions (e.g., *irytacja* 'irritation' (24) and *furia* 'fury' (20))—see Fig. 7.2.

The co-occurrence values between *revulsion* and the British anger cluster emotions are somewhat lower than the co-occurrences between *disgust* and the British anger cluster emotions and somewhat higher than the co-occurrences between *wstręt* 'repulsion, revulsion' and the Polish anger cluster emotions.

The co-occurrence strengths between *wstręt* 'repulsion, revulsion' and the Polish anger cluster emotions are somewhat higher than the interconnections between *obrzydzenie* 'repulsion, disgust, revulsion' and the Polish anger cluster emotions and both of these are much higher than the corresponding interconnections for *zniesmaczenie* 'disgust' (see Fig. 7.3).

Fig. 7.2 Co-occurrence values between wstręt/obrzydzenie and anger and fear clusters

7.3.1.3 British English Versus Polish Disgust—Fear Cluster Co-Occurrences

Comparing British English and Polish co-occurrences in Figs. 7.1, 7.2 and 7.3 it can be seen that the co-occurrences between fear cluster emotions and *wstręt* 'repulsion, revulsion' (e.g., *groza* 'awe, dread, terror' (16), *histeria* 'hysteria' (14) and *panika* 'panic' (14)), *obrzydzenie* 'repulsion, disgust, revulsion' (e.g., *groza* 'awe, dread, terror' (14), *histeria* 'hysteria' (15) and *panika* 'panic' (18)) and *revulsion* (e.g., *dread* (10), *terror* (15), *hysteria* (13) and *panic* (12)) are higher than the corresponding values for both *disgust* (e.g., *dread* (8), *terror* (9), *hysteria* (11) and *panic* (8)) and *zniesmaczenie* 'disgust' (e.g., *groza* 'awe, dread, terror' (8) and *histeria* 'hysteria' (10)). The similar pattern of results for *revulsion* and *obrzydzenie* 'repulsion, disgust, revulsion' reveals that physical disgust is also a salient element of these emotions.

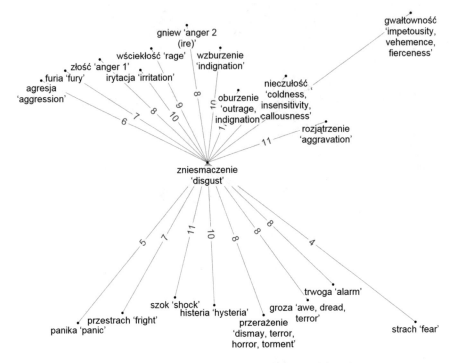

Fig. 7.3 Co-occurrence values between zniesmaczenie and anger and fear clusters

By contrast, *disgust* and *zniesmaczenie* 'disgust' are characterised less by fear and therefore physical disgust is less prominent in these emotions.

7.3.1.4 Conclusions

All in all, compared with the interconnections with the fear cluster emotions, the relatively higher co-occurrences between both the British English and Polish disgust emotions and the anger cluster emotions point to the relatively greater salience of moral disgust. An explanation for this becomes clear when it is considered that moral disgust is at the heart of human values and is elicited by important instances involving, honour, trust, the violation of norm and rights etc.

Further results showed cross-linguistic differences in the different types of disgust in terms of the relative salience of moral versus physical disgust. By virtue of its relatively greater co-occurrences with the anger cluster emotions, *disgust* is characterised more in terms of moral disgust than the other disgust types. The respective co-occurrence values between the anger cluster emotions and *disgust*, *revulsion* and *wstręt* 'repulsion, revulsion', suggest that *revulsion* is characterised by somewhat less moral disgust than *disgust* but somewhat more moral disgust than *wstręt* 'repulsion, revulsion'. The similarity of the lower co-occurrence values between anger cluster emotions and both *wstręt* 'repulsion, revulsion' and *obrzydzenie* 'repulsion, disgust, revulsion' suggests that moral disgust is less salient in both of these emotions, particularly in *obrzydzenie* 'repulsion, disgust, revulsion'. The corresponding low interconnection scores for *zniesmaczenie* 'disgust' were the lowest of the five emotions, possibly indicting very low levels of moral disgust.

The relatively greater element of fear in *wstręt* 'repulsion, revulsion', *revulsion* and *obrzydzenie* 'repulsion, disgust, revulsion' suggests that these types of disgust are characterised relatively more by physical disgust. By contrast, *disgust* and *zniesmaczenie* 'disgust' are characterised less by fear and therefore physical disgust is less prominent in these emotions.

7.3.2   GRID Results

The GRID analyses on disgust are designed to provide a further comparison of the elements of physical and moral disgust in *disgust* and *wstręt* 'repulsion, revulsion'. The conclusion drawn from the online emotions sorting results that *disgust* is characterised more by moral disgust on the basis of its conceptual propinquity with anger means that *disgust* should be related more closely to features of both anger and moral transgression. It is therefore hypothesised that GRID features pertaining to moral transgressions and anger are significantly more likely to occur for *disgust* compared than *wstręt*. On the basis of the online emotions sorting results showing that *wstręt* 'repulsion, revulsion' is characterised relatively more in terms of physical disgust due to its conceptual proximity with fear, it is additionally hypothesised that GRID features relating to fear are significantly more likely to occur for *wstręt* 'repulsion, revulsion' than *disgust*. As the usual experience of fear when confronted with a revolting, physically disgusting situation is likely to often be accompanied by withdrawal, it can be further predicted that the GRID features relevant to withdrawal will also be more likely to occur for *wstręt* 'repulsion, revulsion' than *disgust*.

Different sources were consulted in the selection of GRID features to be included in the present study. All of the features present in Lee and Ellsworth [18] were used as either *withdrawal/fear* features or *moral transgression/anger* features. In addition, the features that Soriano [30] identified *as desire to act* features pertaining to anger were included in the *moral transgression/anger* features. Also, the FRIGHT/low power features that were used in our analyses of fear [19] were added to the *withdrawal/fear* features. There were 25 *withdrawal/fear* features and 23 *moral transgression/anger* features (see Table 7.1).

The means of the *withdrawal/fear* features and the *moral transgression/anger* features pertaining to disgust were determined for each participant. A 2×2 Anova was performed on these means that had one between-subjects variable (language group: British English *disgust* versus Polish *wstręt*). There was also one within-subjects variable (GRID features: *withdrawal/fear* features versus *moral transgression/anger* features—see Table 7.1). There was a significant main effect of language group, $F(1, 57) = 4.83$, p < 0.05. British English *disgust* was associated with a higher

Table 7.1 GRID features characterised by withdrawal/fear and moral transgression/anger

Withdrawal/fear features	Moral transgression/anger features
Caused by chance	Incongruent with own standards and ideals
Felt weak limbs	Violated laws or socially accepted norms
Had a lump in the throat	Treated unjustly
Decreased the volume of voice	Caused by somebody else's behaviour
Had a trembling voice	Required an immediate response
Fell silent	Consequences avoidable or modifiable
Spoke slower	Moved against people or things
Felt the urge to stop what he or she was doing	Increased the volume of voice
Wanted to undo what was happening	Had an assertive voice
Wanted to comply with someone else's wishes	Spoke faster
Wanted to hand over the initiative to someone else	Wanted to go on with what he or she was doing
Wanted to submit to the situation as it is	Wanted to be in control of the situation
Wanted someone to be there to provide help and support	Wanted to take the initiative him/herself
Lacked the motivation to do anything	Felt urge to be active, to do something, anything
Wanted to do nothing	Wanted to move
Lacked motivation to pay attention to what was going on	Felt urge to be attentive to what is going on
Wanted to prevent or stop sensory contact	Wanted to do damage, hit, or say something that hurts
Wanted to disappear or hide from others	
Wanted to make up for what he or she had done	Wanted to oppose
Wanted to run away in whatever direction	Wanted to tackle the situation
Felt submissive	Wanted to destroy whatever was close
Felt powerless	Wanted to act, whatever action might be
Felt negative	Felt powerful
Felt bad	Felt dominant
Felt weak	

likelihood of occurrence on both sets of GRID features than Polish *wstręt* (means of 5.75 and 5.3, respectively). There was also a significant interaction between language group and GRID features, $F(1, 57) = 26.06$, p < 0.001. Contrasts were performed to beak down this interaction. There was a significant difference between *disgust* and *wstręt* on the *moral transgression/anger* features, $F(1, 57) = 20.41$, p < 0.001. Figure 7.4 shows that the *moral transgression/anger* features are relatively more salient for *disgust* than *wstręt* (means of 6.0 and 4.82, respectively). By contrast, there was no significant difference between *disgust* and *wstręt* on *withdrawal/fear* features, $F(1, 57) = 1.58$, p > 0.05. *Disgust* had significantly higher scores on the

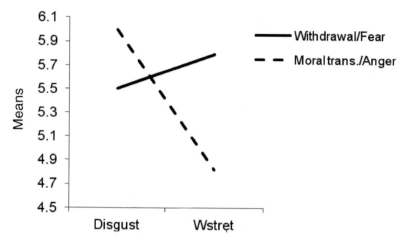

Fig. 7.4 Means of disgust versus wstręt on *withdrawal/fear* features versus *moral transgression/anger* features

moral transgression/anger features (mean = 6.0) than the *withdrawal/fear* features (mean = 5.5), $F(1, 57) = 4.24$, p < 0.05. *Wstręt* had significantly higher scores on the *withdrawal/fear* features (mean = 5.79) than *moral transgression/anger* features (mean = 4.82), $F(1, 57) = 19.76$, p < 0.001.

7.3.2.1   GRID Conclusions

The data suggest that whereas *wstręt* is characterised relatively more by *withdrawal/fear*, *moral transgression/anger* is relatively more salient in *disgust*. When comparing *disgust* with *wstręt*, the results show that the *moral transgression/anger* features were rated higher on *disgust* than *wstręt*. However, there was no significant difference between the two emotions on *withdrawal/fear* features. It can be concluded that whereas *withdrawal/fear* is relatively more of a feature of *wstręt*, *moral transgression/anger* is more of a salient element of *disgust*, and that the hypotheses are therefore supported.

7.3.3   Corpus Results

7.3.3.1   Somatic and Bodily Symptoms of Disgust Expression

The words that are semantically related to disgust in physiological terms are vomit and excrement, which is the reason why a person feeling disgust expressively displays it as a *shudder of disgust* in English *and as dreszcz obrzydzenia* in Polish. Bodily symptoms of disgust can be identified both as facial expressions (moue, grimace, look of disgust), body movements (recoil) gastric processes (vomit, in the pit of the stomach) and language (shouting exclamations of disgust), as is evidenced in the monolingual and translation corpus data:

Expressiveness (bodily symptoms)—usually with reference to physical disgust (repulsion)

(14) A sort of vomit of **disgust**
(15) Burden made a moue of **disgust**
(16) though not without a grimace of **disgust**, did we swallow it
(17) He recoiled in apparent **disgust**
(18) shouting exclamations of **disgust**
(19) this has always been a constant, the feeling of **disgust** in the pit of the stomach
(20) reached into the bin with a look of **disgust** on his face
(21) Pol. **strach i obrzydzenie** zatamowały   Eng. but **fear and aversion** restrained
     mu jednak głos w piersiach   the voice in his bosom
(22) Pol. **Wstręt, bezgraniczny, mdlący**   Eng. His **indefinite nausea** would not
     wstręt [lit. 'unbounded, sicken-   let him stay.
     ing repulsion] nie pozwalał mu się
     zatrzymać.
(23) Pol. A nie **odpychał od niego**   Eng. instead of **shrinking from him**
     najgłębszy wstręt   **with the strongest repugnance**
(24) Pol. Chciał tego Loerke, ale pow-   Eng. He wanted it, but was **held back**
     strzymywał go **nieodparty wstręt**   **by some inevitable reluctance**.

The somatic and behavioural responses in a disgust event are similar cross—linguistically; what is different though is the intensity, degree of arousal and the unpleasantness judgement in such cases.

7.3.3.2   Elements of DISGUST Scenarios

The acquisition of data from large language corpora makes it possible to build pro-totypical Emotion Event scenarios. Properties of the emotions expressed by means of Adjectival collocates, typical characteristics of emotion expressions displayed by relevant Verbs, as well as Noun collocates of the words derived from particular emotion terms, e.g., *disgusting*, are the materials which provide, as exemplified below, relevant information on the scenarios of emotions studied.

7.3.3.3   *Disgust* and Its Objects

We compare textual collocates, i.e., words which co-occur in corpora more fre-quently than by chance, for a number of words used in the expression or descrip-tion of *wstręt/disgust* scenarios in Polish and English. A Colosaurus comparison (http://pelcra.pl/hask_en/GraphGenerator)[3] between the Adj *disgusting, repulsive*

---

[3]Colosaurus and the collate generator are the tools developed by Piotr Pęzik [24]. Colosaurus is used for comparing the textual collocates of selected words.

and *revulsive* and Noun combinations below shows both their frequency pattern with the adjective *disgusting* to be used most frequently and in the widest range of contexts. The exception is the collocate with the Noun *force*, which is used uniquely with the attribute *repulsive* in the field of physics. The adjective *revulsive* is not listed in any collocate combination as it does not exceed the frequency of 3 or above.

List 1: Colosaurus *disgusting (d)–repulsive (r)*

Collocate	Count	d_A	d_TTEST	d_R	r_A	r_TTEST	r_R
behaviour	1	3	0.73	2	0	–	0
book	1	3	−1.15	3	0	–	0
creature	1	3	1.43	3	0	–	0
display	1	4	1.63	3	0	–	0
force	1	0	–	0	14	3.57	11
habit	1	4	1.74	3	0	–	0
language	1	3	0.0	3	0	–	0
object	1	3	0.97	3	0	–	0
thing	1	16	1.43	13	0	–	0
way	1	7	−2.91	7	0	–	0

Objects of *wstręt* in Polish

List 2: *obrzydliwy* (derivative of *brzydzić się* 'to shudder, shrink from something ugly or dirty')

#	Collocate	POS	A^a	TTEST	MI3	
1	coś	noun	62.0	6.66	14.61	'something'
2	smak	noun	23.0	4.68	14.51	'taste'
3	zapach	noun	13.0	3.41	11.61	'smell'
4	rzecz	noun	19.0	2.85	10.03	'thing'
5	ruda	noun	9.0	2.84	10.57	'red-haired'
6	pomówienie	noun	8.0	2.80	12.68	'slander'
7	smród	noun	7.0	2.60	11.51	'stink'
8	fetor	noun	6.0	2.44	13.41	'stench'
9	ciało	noun	10.0	2.29	8.51	'body'
10	czyn	noun	7.0	2.26	8.41	'deed'
11	postać	noun	10.0	2.25	8.43	'figure'
12	pająk	noun	5.0	2.20	10.67	'spider'
13	kawa	noun	6.0	2.16	8.27	'coffee'
14	manipulacja	noun	5.0	2.14	9.33	'manipulation'
15	plotka	noun	5.0	2.10	8.78	'gossip'

[a]The symbol "A" denotes RAW FREQUENCIES of COLLOCATES, i.e., words typically co-occurring with other ones with a frequency greater than chance

List 3: *wstrętny* **(derivative of** *wstręt* **'repulsion')**

#	Collocate	POS	A	TTEST	MI3	
1	coś	Noun	30.0	4.24	11.97	'something'
2	rzecz	Noun	20.0	3.43	10.75	'thing'
3	baba	Noun	12.0	3.38	12.68	'woman' (offensive)
4	zapach	Noun	12.0	3.32	11.76	'smell'
5	bachor	Noun	8.0	2.82	15.00	'child, brat' (offensive)
6	odór	Noun	8.0	2.81	14.13	'odour'
7	babsko	Noun	7.0	2.64	14.52	'woman' (offensive)
8	okropna	Noun	7.0	2.61	11.92	'terrible'
9	typ	Noun	10.0	2.54	8.99	'type'
10	smak	Noun	7.0	2.50	9.86	'taste'
11	dziad	Noun	6.0	2.40	11.09	'old beggar' (offensive)
12	chłopak	Noun	7.0	2.32	8.65	'boy'
13	plotka	Noun	5.0	2.14	9.27	'gossip'
14	obca	Noun	5.0	2.09	8.60	'foreign'
15	ptaszysko	Noun	4.0	1.99	12.58	'bird' (augmentative)
16	egoista	Noun	4.0	1.99	11.66	'egoist'

The collocates are mostly Nouns referring to the sense of smell on the one hand or to augmentative suffixed-Nouns that are offensive or vulgar; in the list there are three offensive forms addressed to females.

The physical sense of *wstręt* can also be caused by *dirt* as evidenced in the Polish translation data in which the original *dirty* is rendered as *causing repulsion*:

(25) A czy zawsze **budził w pani wstręt?** "And **was he always dirty?**" he asked.
   lit. 'did he always wake up (metaphoric)
   repulsion in you?'

List 4: *niesmaczny* (derivative of *niesmak* 'dis+taste', lexically related to *zniesmacznie*, used with reference to physical and abstract objects)

#	Collocate	POS	A	TTEST	MI3	
1	żart	Noun	23.0	4.76	16.49	'joke'
2	coś	Noun	15.0	3.14	10.22	'something'
3	dowcip	Noun	8.0	2.80	12.76	'joke'
4	scena	Noun	8.0	2.62	9.80	'scene'
5	jedzenie	Noun	6.0	2.37	10.27	'food'
6	spektakl	Noun	5.0	2.15	9.43	'spectacle, performance'
7	mięso	Noun	5.0	2.15	9.35	'meat'
8	farsa	Noun	3.0	1.72	11.23	'farce'
9	zjedzenie	Noun	3.0	1.72	10.88	'eating'
10	troszeczkę	Noun	3.0	1.71	10.13	'somewhat'
11	przesada	Noun	3.0	1.69	8.91	'exaggeration'
12	potrawa	Noun	3.0	1.68	8.30	'dish'
13	danie	Noun	3.0	1.67	8.18	'dish'
14	piwo	Noun	3.0	1.58	6.73	'beer'
15	reklama	Noun	3.0	1.52	6.24	'advertisement'

It is only with the adjective *niesmaczny* 'dist-tasteful' that the most frequent collocate refers to a moral judgment (*niesmaczny żart/dowcip* 'disgusting joke'), but starting from the object identified in (5), the collocates are physical objects, usually food or drink. The concept of *niesmak* (lit. 'non-taste, dis-taste') can be understood either as a physical sensation or moral distaste (10). In the formal language of official documents, it is an equivalent of Eng. *distaste* or *aftertaste* respectively, much less frequently—*disgust*.

List 5: Polish *wstręt* 'repulsion, revulsion'

Adjectival collocates

#	Collocate	POS	A	TTEST	MI3	
1	wrodzony	adj	12.0	3.43	14.12	'inborn'
2	fizyczny	adj	12.0	3.08	10.35	'physical'
3	wszelki	adj	13.0	3.01	10.01	'any'
4	nieprzezwyciężony	adj	9.0	2.99	17.40	'invincible'
5	wyraźny	adj	8.0	2.61	9.75	'clear'
6	pełny	adj	11.0	2.42	8.81	'full'
7	nagły	adj	6.0	2.29	9.13	'sudden'
8	niepohamowany	adj	5.0	2.22	12.39	'uncontrollable'
9	głęboki	adj	6.0	2.09	7.96	'deep'
10	powszechny	adj	6.0	1.99	7.60	'common'
11	nieprzeparty	adj	4.0	1.99	12.88	'uncontrollable'
12	sam	adj	27.0	1.80	10.12	'only'
13	nieprzełamany	adj	3.0	1.73	15.37	'unbreakable'
14	niekłamany	adj	3.0	1.72	10.69	'sincere'
15	straszny	adj	4.0	1.72	6.84	'terrible'
16	heteroseksualny	adj	3.0	1.72	10.44	'heterosexual'
17	pomieszany	adj	3.0	1.71	9.97	'mixed'
18	organiczny	adj	3.0	1.67	8.03	'organic'
19	przejęty	adj	3.0	1.63	7.39	'filled with/taken over by'
20	zwyczajny	adj	3.0	1.57	6.65	'common'
21	zimny	adj	3.0	1.56	6.53	'cold'

A characteristic property distinguishing between *obrzydzenie* and *wstręt* is the fact that *wstręt* is considered a stronger emotion, judging by a significantly higher number of synonymic Adjectival collocates of *wstręt* used in the sense of 'uncontrollable' such as the metaphoric conceptualisations in terms of invincible (*nieprzezwyciężony*), impossible to stop (*niepohamowany*), unbreakable, unbanned (*nieprzełamany*), irrefutable (*nieodparty*) or lit. without lying, true, real (*niekłamany*), all absent from the *obrzydzenie* combinations. The metaphoric conceptualisations of *wstręt* present a range of Source Domains relating to the *enemy* frame (winning scenario), *engine* (with the brakes to stop the emotion), argument (not to refute) or *truth*. It is also *wstręt*, and not *obrzydzenie*, that can be considered *cold, organic* or *deep*. The verbal collocates are semantically closer to each other and point to bodily reactions connected with disgust such as abhorring, shudder, or twist(ed) (grimaces).

English *disgust* has more elements of Polish *obrzydzenie* than of *wstręt* but more clearly shows some properties combined with *contempt, indignation*, and *rejection* (unacceptance). Polish *wstręt* and *obrzydzenie* are members of the same cluster. The basic prototypical sense of both is more frequently related to a physical aversion and withdrawal than in the case of English *disgust*. The construal of the DISGUST event is not identical across English and Polish. The list of Verbal collocates below makes it possible to build a prototypical wstręt scenario in Polish:

List 6: Pol. *wstręt* 'repulsion, revulsion'
Verbal collocates

#	Collocate	POS	A	TTEST	MI3	
1	czuć	verb	108.0	10.19	19.21	'feel'
2	budzić	verb	90.0	9.41	19.94	'wake up' (metaphoric)
3	mieć	verb	120.0	7.12	15.33	'have'
4	napawać	verb	28.0	5.28	18.98	'fill with'
5	odczuwać	verb	28.0	5.23	16.17	'feel'
6	czynić	verb	21.0	4.45	13.94	'do'
7	odrzucić	verb	18.0	4.14	13.76	'reject'
8	odwracać	verb	17.0	4.09	15.14	'turn away'
9	poczuć	verb	17.0	3.98	13.05	'start feeling'
10	nabrać	verb	15.0	3.81	13.95	'take'
11	wywoływać	verb	15.0	3.78	13.31	'call forth'
12	patrzyć	verb	16.0	3.56	11.20	'look'
13	żywić	verb	12.0	3.42	13.75	'feed' (metaphoric) 'feel'
14	wzbudzać	verb	12.0	3.41	13.42	'wake'
15	przejmować	verb	11.0	3.24	12.36	'take from'
16	ogarnąć	verb	10.0	3.11	12.77	'overwhelm' perf
17	ogarniać	verb	9.0	2.96	12.97	'overwhelm'
18	odsunąć	verb	9.0	2.95	12.48	'move away'
19	napełniać	verb	8.0	2.81	13.56	'fill with'
20	otrząsnąć	verb	7.0	2.63	13.36	'shake off'

List 7: Pol. *obrzydzenie* (derivative of *brzydki* 'ugly (caused by ugliness)'
Verbal collocates

#	Collocate	POS	A	TTEST	MI3	
1	czuć	verb	61.0	7.54	16.72	'feel'
2	budzić	verb	56.0	7.38	17.87	'wake up'
3	napawać	verb	38.0	6.15	20.29	'fill with'
4	patrzyć	verb	35.0	5.61	14.57	'look'
5	splunąć	verb	22.0	4.68	18.92	'spit'
6	wzbudzać	verb	19.0	4.32	15.40	'wake'
7	brać	verb	19.0	3.90	11.75	'take'
8	poczuć	verb	14.0	3.58	12.20	'start feeling'
9	nabrać	verb	13.0	3.54	13.31	'take'
10	spojrzeć	verb	14.0	3.54	11.84	'look'
11	wywoływać	verb	13.0	3.51	12.68	'call forth'
12	wywołać	verb	13.0	3.47	12.17	'call forth' perf
13	krzywić	verb	12.0	3.45	15.99	'twist'
14	skrzywić	verb	12.0	3.45	15.13	'twist' perf
15	ogarniać	verb	12.0	3.43	14.20	'overwhelm'
16	wzdrygnąć	verb	11.0	3.31	16.05	'wince, abhor'
17	otrząsnąć	verb	11.0	3.30	15.30	'shake off'
18	odczuwać	verb	10.0	3.06	11.70	'feel'
19	wstrząsać	verb	9.0	2.99	14.91	'shake, shudder'

The results of the corpus-based analysis of English *disgust* and Polish *wstręt* show that the meaning of *wstręt* overlaps to a great extent with that of *'repulsion, revulsion'* emotions, characterised, similarly to *repulsion/revulsion*, by greater physical disgust and its manifestations in shrinking, sickness, etc.

A significantly higher number of synonymic Adjectival collocates of *wstręt* with the meaning of 'uncontrollable' (invincible *nieprzezwyciężony*, impossible to stop *niepohamowany*, unbreakable *nieprzełamany*, irrefutable), which are not identified for *obrzydzenie* combinations, points to the experiencer's higher arousal in the case of *wstręt* when compared with *obrzydzenie*. The verbal collocates of *wstręt* indicate bodily reactions that are also connected with disgust such as abhorring, shuddering, and facial grimaces.

7.3.3.4 Translational Data

The semantic overlap of co-occurrence values of *wstręt* with other negative emotions such as hate, anger, and fear is evident in the translational Polish-to-English and English-to-Polish data.

DISGUST Cluster *Wstręt/Disgust*
Disgust and fear

(26) Mam **wstręt [lit. repulsion/disgust] do** I have quite a ***horror of upstarts***.
parweniuszów
(27) Trzeba panu wiedzieć, że nie cierpię fir- You must know I have a vast dislike to
cyków, **mam po prostu wstręt** do nich. puppies—quite a **horror** of them

Disgust and hate

(28) Kłamstwo w pani ustach wzbudziłoby I should almost have **hated** you, had you
we mnie **wstręt [lit. your lying would** flattered and lied
arouse disgust/repulsion in me].
(29) poczułem taki wstręt i taką nien- I was supported by a **scornful detestation**
awiść **[lit. repulsion and hate]**, że of him that sealed my lips
zapieczętowały mi one usta.

Disgust is an emotion which is most frequently expressed in language as either the property of the emotion itself, its bodily or mental reaction or else as another emotion, which is causally linked to *disgust*. In numerous other contexts a number of other English *disgust* cluster related emotion concepts function as equivalents of Polish *wstręt*. Polish *wstręt* is semantically related with the physical object more frequently than English *disgust*, which is linked to a number of intracluster members, used most frequently with morally unacceptable objects and events.

Polish   *wstręt* lit. 'repulsion', *brzydzić się* lit. 'to find something dirty and to abhor because of that', *wzdrygnąć/wzdrygać się* lit. 'shrink from something'

English   *disgust, aversion, nausea, distaste, horror, despise, loathe, shrinking (from something), execrate, repugnance, hate, speech becomes insipid, abominance, to be put upon somebody, to be sick of sb/smth, scorn, seldom have a pleasure, scornful detestation, dislike, revulsion, antipathy, animosity, reluctance, bad feeling, nausea*

Individual members of English *disgust* and Polish *wstręt* clusters have a number of common properties, particularly with reference to physical objects of *disgust/wstręt* and physical, somatic and bodily manifestations of these emotions. And yet, they are not full equivalents when considered in terms of individual one-to-one correspondences. Rather, they can be taken to act as corresponding clusters, which, as will be additionally argued below, display culture/language-bound inter-cluster associations.[4]

---

[4]The consultation of the additional semantic resources, where common semantic properties and relations between the two languages are presented, such as the 'synsets' in English and Polish Wordnet, does not reveal additional information at the investigated conceptual level: Eng. (Noun) *disgust* (strong feelings of dislike), Verb: *disgust, gross out, revolt, repel* (fill with *distaste*; *disgust, revolt, nauseate, sicken, churn up* (cause aversion in; offend the moral sense of); http://wordnetweb.princeton.edu/perl/webwn?c=0\&sub=Change\&o2=\&o0=1\&o8=

7.4   Conclusions

The results of the online emotions sorting study showing that disgust in British English and Polish is characterised more by anger than fear, points to the greater salience of moral disgust in conceptualisations of this emotion in both cultures. As moral disgust is central to our code of values, and hence an integral part of the essence of humanity, it is rather obvious that it is more salient than physical disgust. The only possible exception to this pattern, although this is not confirmed in the corpus data, is *zniesmaczenie* 'disgust', which has more of a relative balance in terms of moral and physical disgust.

Overall, the corpus, GRID and online emotions sorting methodologies produced generally consistent results regarding the comparison of British English and Polish types of disgust. Consistent results showing that *wstręt* 'repulsion, revulsion' has higher ratings on the *withdrawal/fear* GRID features and has relatively high co-occurrence values with fear cluster emotions supports the hypothesis that this emotion is characterised by relatively greater physical disgust. The salience of physical disgust in *obrzydzenie* 'repulsion, disgust, revulsion' is demonstrated by the similarity it shares with *wstręt* 'repulsion, revulsion' in terms of its relatively high interconnections with the fear cluster emotions, compared with the lower corresponding co-occurrences for *disgust*. This pattern is consistent with the corpus results showing that both *wstręt* 'repulsion, revulsion' and *obrzydzenie* 'repulsion, disgust, revulsion' have more of a sense of physical aversion and withdrawal than *disgust*. As one would expect of emotions related more to physical disgust, *wstręt* 'repulsion, revulsion' and *obrzydzenie* 'repulsion, disgust, revulsion' have similar interconnections with Polish fear cluster emotions as *revulsion* does to its respective British English fear cluster emotions.

The relatively greater element of moral disgust in *disgust* is seen by the higher ratings for this emotion than *wstręt* 'repulsion, revulsion' on *moral transgression/anger* GRID features and the relatively high co-occurrence values with the anger cluster emotions. The corpus results similarly show that *disgust* is similarly linked with more moral aspects than both *wstręt* 'repulsion, revulsion' and *obrzydzenie* 'repulsion, disgust, revulsion', which can be seen in more of its properties related to contempt, indignation and rejection. *Obrzydzenie* 'repulsion, disgust, revulsion' shows slightly lower co-occurrence values with anger cluster emotions than *wstręt* 'repulsion, revulsion' and therefore suggests that moral disgust is also less characteristic of this emotion. The corresponding low interconnection scores for *zniesmaczenie* 'disgust' indicates very low levels of moral disgust. In comparison, the interconnections between *revulsion* and the anger cluster emotions suggest that compared with *disgust*

1\&o1=1\&o7=\&o5=\&o9=\&o6=\&o3=\&o4=\&i=-1\&h=000\&s=disgust. Pol *awersja, repulsja, obmierzłość, odraza, obrzydzenie, (obrzydliwość)* http://plwordnet.pwr.wroc.pl/wordnet/3efcdb76-c0f8-11e4-ac52-7a5d273e87eb. It might be interesting, in a separate study, to compare the results we also gained from the online study with the location of these forms in the database and with more extended *disgust/wstręt* elaboration paths in these resources.

and *wstręt* 'repulsion, revulsion', *revulsion* is characterised by an intermediate level of moral disgust.

Robots might display what would be considered disgust behaviour, building representations of emotions on the basis of sensory input data and knowledge acquired from large storages of data. They are not able to *replicate* human cognitive and emotional mechanisms at present [31, 33], but can recognise and signal intentions indirectly by means of bodily movements and gestures, which makes the task similar to producing and reading affective states. They can learn from the information implicit in emotions, and recognise the emotional states of human interactants, augmented by the knowledge acquired from large corpus data and real-life contexts. However, in the light of the developing field of machine ethics the mere encoding and decoding of disgust might in the future be secondary to a more fundamental role of this emotion. As Anderson and Anderson [1] explain, "of the many challenges facing those who choose to work in the area of machine ethics, foremost is the need for a dialogue between ethicists and researchers in artificial intelligence" (p. 25). Given the important element of ethics and morality in disgust, it is clear that this emotion is central to the development of ethical robots.

While for physical disgust it is sufficient to have a number of basically external causes/stimuli (important corpus generated lists) and external (bodily reactions) with basic lexical/prosodic linguistic patterns identified, moral disgust requires more sophisticated conditioning and involves more extensive cognitive representations. This study focuses mainly on disgust but has relevance to other social and moral emotions. Although the moral emotion megacluster *disgust* is a member of and shares properties with the 'other-condemning' emotions, *anger* and *contempt* [11], it is distinct from them on the basis of *physical* stimulus conditioning, behavioural reactions and effects. Self-conscious emotions on the other hand, particularly *guilt* and *regret*, enter a distinct interactional pattern than the *disgust* cluster, with *guilt* being more internally directed than either *shame* or *disgust* [20]. Other moral emotion clusters like *compassion* and *gratitude* display a more socially oriented, interactional character, and are similar to *disgust* in their moral manifestation, although they are distinct on their polarity marking, which is positive in their case and associated with different stimulus objects and conditioning. Their behavioural and facial properties as well as linguistic expression also build a profile that is specific for particular emotions. Cross-cultural properties of disgust, exemplified by English and Polish in the present paper, also involve systematic external differences in their manifestation, particularly with reference to their stimuli and exbodiment, i.e., enactment of particular cultures [35], which are potentially recognisable for socially believable systems in affective robotics.

References

1. Anderson M, Anderson LW (2007) Machine ethics: creating an ethical intelligent agent. ai mag 28(4):15–26
2. Breazeal CL (2004) Designing sociable robots. MIT Press, Cambridge

3. Brinkmann S, Musaeus P (2012) Emotions and the moral order. In: Wilson PA (ed) Dynamicity in emotion concepts. Lodz studies in language, vol 27. Peter Lang, Frankfurt a. Main, pp 123–137

4. Cañamero L, Gaussier Ph (2005) Emotion understanding: Robots as tools and models. In: Muir D (ed) Nadel J. Emotional development, Recent research advances. OUP, pp 235–258

5. Cushman F, Gray K, Gaffey A et al (2012) Simulating murder: the aversion to harmful action. Emotion 12(1):2–7

6. Dubé L, Le Bel J (2003) The content and structure of laypeople's concept of pleasure. Cognit Emot 17(2):263–295

7. Ellsworth PC, Scherer KR (2003) Appraisal processes in emotion. In: Davidson RJ, Scherer KR, Goldsmith H (eds) Handbook of affective sciences. Oxford University Press, New York, pp 572–595

8. Fontaine JJR, Poortinga YH, Setiadi B et al (2002) Cognitive structure of emotion terms in Indonesia and The Netherlands. Cognit Emot 16(1):61–86

9. Fontaine JJR, Scherer KR, Soriano C (2013) Components of emotional meaning: a sourcebook. Oxford University Press, Oxford

10. Gutierrez R, Giner-Sorolla R, Vasiljevic M (2012) Just an anger synonym? Moral context influences predictors of disgust word use. Cognit Emot 26(1):53–64

11. Haidt J (2003) The moral emotions. In: Davidson RJ, Scherer KR, Goldsmith H (eds) Handbook of affective sciences. Oxford University Press, New York, pp 852–870

12. Heider KG (1991) Landscapes of emotion: mapping three cultures of emotion in Indonesia. Cambridge University Press, Cambridge

13. Hudlicka E, McNeese MD (2002) Assessment of user affective and belief states for interface adaptation: application to an air force pilot task. User Model User-Adapt Inter 12(1):1–47

14. Jones D (2007) The depths of disgust. Nature 447:768–771

15. Lakoff G, Johnson M (1980) Metaphors we live by. Chicago University Press, Chicago

16. Lakoff G, Espenson J, Goldberg A et al (1991) Master metaphor list, second draft copy. Cognitive Linguistics Group, Univeristy of California, Berkeley

17. Langacker RW (1987, 1991) Foundations of cognitive grammars, vol I, vol II. Stanford University Press, Stanford

18. Lee SWS, Ellsworth PC (2013) Maggots and morals: physical disgust is to fear as moral disgust is to anger. In: Fontaine JJR, Scherer KR, Soriano C (eds) Components of emotional meaning: a sourcebook. Oxford University Press, Oxford, pp 271–280

19. Lewandowska-Tomaszczyk B, Wilson PA (2013) English fear and Polish strach in contrast: GRID approach and cognitive corpus linguistic methodology. In: Fontaine JJR, Scherer KR, Soriano C (eds) Components of emotional meaning: a sourcebook. Oxford University Press, Oxford, pp 425–436

20. Lewandowska-Tomaszczyk B, Wilson PA (2014) Self-conscious emotions in collectivistic and individualistic cultures: a contrastive linguistic perspective. In: Romero-Trillo J (ed) Yearbook of corpus linguistics and pragmatics, vol 2: New empirical and theoretical paradigms. Springer, Berlin, pp 123–148

21. Moll J, de Oliveira-Souza R, Moll FT et al (2005) The moral affiliations of disgust: a functional MRI study. Cognit Behav Neurol 18:68–78

22. Niedenthal PM, Krauth-Gruber S, Ric F (2006) The psychology of emotion: interpersonal, experiential, and cognitive approaches. Psychology Press, New York

23. Ortony A, Clore GL, Collins A (1988) The cognitive structure of emotions. Cambridge University Press, Cambridge

24. Pezik P (2014) Graph-based analysis of collocational profiles. In: Jesenšek V, Grzybek P (eds) Phraseologie im wörterbuch und korpus (phraseology in dictionaries and corpora). ZORA 97. Maribor, BielskoBiała, Budapest, Kansas, Praha, Filozofska fakuteta

25. Reips U-D (2002) Standards for internet-based experimenting. Exp Psychol 49:243–256

26. Royzman EB, Sabini J (2001) Something it takes to be an emotion: the interesting case of disgust. J Theory Soc Behav 31:29–59

27. Scherer KR (2005) What are emotions? And how can they be measured? Soc Sci Inf 44:693–727

28. Schnall S, Benton J, Harvey S (2008) With a clean conscience: cleanliness reduces the severity of moral judgements. Psychol Sci 19(12):1219–1222
29. Smith M, Ceni A, Milic-Frayling, N, Shneiderman, B, Mendes Rodrigues, E, Leskovec, J, Dunne, C (2010) NodeXL: A free and open network overview, discovery and exploration add-in for Excel 2007/2010/2013/2016, http://nodexl.codeplex.com/ from the Social Media Research Foundation, http://www.smrfoundation.org
30. Soriano C (2013) Conceptual metaphor theory and the GRID paradigm in the study of anger in English and Spanish. In: Fontaine JJR, Scherer KR, Soriano C (eds) Components of emotional meaning: a sourcebook. Oxford University Press, Oxford, pp 410–424
31. Spinola de Freitas J, Queiroz J (2007) Artificial emotions: are we ready for them? In: Almeida e Costa F, Rocha LM, Costa E, Harvey I, Coutinho A (eds) ECAL 2007, LNAI, vol 4648. Springer, Berlin, pp 223–232
32. Vogel C (2014) Denoting offence. Cognit Comput 6(4):628–639
33. Wallach W, Allen C (2009) Moral machines: teaching robots from right and wrong. Oxford University Press, New York
34. Wilson PA, Lewandowska-Tomaszczyk B (2014) Affective robotics: modelling and testing cultural prototypes. Cognit Comput 6(4):814–840
35. Zatti A, Zarbo C (2015) Embodied and exbodied mind in clinical psychology. A proposal for a psycho-social interpretation of mental disorders. Front Psychol 6:1–7. doi:10.3389/fpsyg. 2015.00236

Chapter 8
Speaker's Hand Gestures Can Modulate Receiver's Negative Reactions to a Disagreeable Verbal Message

Fridanna Maricchiolo, Augusto Gnisci, Mariangela Cerasuolo, Gianluca Ficca and Marino Bonaiuto

Abstract Here we report the results of an experiment aimed to investigate the effects of different hand gestures on emotional and attitudinal reactions of receivers through the measure of physiological indexes (facial muscles activity, heart rate, and eye blinking). A videotape was shown to 50 University students, in which an actress presented a speech with a disagreeable verbal content, namely the proposal of increasing University fees. The verbal message included the presentation of four arguments (two strong and two weak) in support of the proposal. During her speech, the actress manipulated "Gesture Type" in order to achieve five conditions (ideational gestures, discursive gestures, object- and self-adaptors, and no gesture as control). ANOVAs reveal that the different type of gestures differently modulate the negative impact of disagreeable verbal content on receivers in different moments of the speech and interact with strong or weak arguments in determining negative reactions to the disagreeable verbal message. In particular, it seems that discursive (conversational and ideational) gestures are more capable to counteract the negative effect of the arguments. These results give a further contribution to a better understanding of the crucial role of gestures in providing characteristics of speech perception, also in terms of persuasion.

F. Maricchiolo (✉)
Department of Education, University of Roma Tre, Rome, Italy
e-mail: fridanna.maricchiolo@uniroma3.it

A. Gnisci · M. Cerasuolo · G. Ficca
Department of Psychology, Second University of Naples, Caserta, Italy
e-mail: augusto.gnisci@unina2.it

M. Cerasuolo
e-mail: mariangela.cerasuolo@gmail.com

G. Ficca
e-mail: gianluca.ficca@unina2.it

M. Bonaiuto
Sapienza University of Rome, Rome, Italy
e-mail: marino.bonaiuto@uniroma1.it

© Springer International Publishing Switzerland 2016
A. Esposito and L.C. Jain (eds.), *Toward Robotic Socially Believable Behaving Systems - Volume I*, Intelligent Systems Reference Library 105,
DOI 10.1007/978-3-319-31056-5_8

Keywords Hand gestures · Psychosocial perception · Preliminary study · Rhythmic gestures · Cohesive gestures · Self-adaptors · Frequency

8.1 Introduction

Oral communication is based on two main channels: verbal and non-verbal (gestures). It is known that hand gestures are crucial elements of verbal communication, since they help the speaker in language planning and production [1–3] as well as in conveying his/her emotions or cognitive effort [4]. Co-speech gestures are an integral part of the dialogue [5], in that they add a significant amount of information to a speaker's message (e.g., [6–8]) and provide communicative advantages for both the speaker and the listener, offering a shared ground for an effective and successful interaction [9]. For instance, it has been shown that co-speech gestures help non-native language learners by letting them better comprehend unfamiliar parts of the spoken message [10]. According to this research line, a recent study on neural activity during conversation shows that perceiving hand movements during speech modulates the neural activation in listeners, specifically involving biological circuits for both motion perception and verbal comprehension [11]. These findings clearly indicate that listeners seek to find meaning not only in the words they hear but also in the hand movements they see [12].

Since hand gestures can serve a better comprehension of the speaker's message [13], they are likely to increase or decrease the impact of a verbal message on the receiver. Listeners have to monitor two different—but integrated—sources of information from the speaker: verbal content and gesture. Different co-speech gestures are differently perceived [14, 15] and differently affect receivers' evaluations of the speaker and the conveyed message [16].

For what said so far, a clear understanding of hand gestures' implications might contribute to implement a coded system of this kind of verbal-analogic interaction: such a system would definitely facilitate "intelligent machines" to better detect the overall meaning of human speech.

Co-speech hand gestures are generally distinguished in different categories. Partly inspired by traditional McNeill's classification system [3] and Kendon's further proposals [2], our group has recently developed a gestures' taxonomic system, according to which it is possible to distinguish: (a) illustrators or ideational gestures, related to the semantic content of concurrent speech and aimed to describe and illustrate it; (b) conversational gestures, accompanying the prosodic of speech, aimed to mark discourse fluidity or cohesion, [3, 17] and without a relation to the semantic content. (c) adaptors, which include self- person-, and object-addressed movements, which are not related to speech, but are mainly directed to manage emotional states, such as tension or anxiety [18].

The majority of the authors agree on the fact that different kinds of gestures may exert different effects on the perceivers according to their specific features. It has been proposed that illustrators improve the attention [20], the understanding accuracy [6] and the recognition and recall about the content described by the sender (speech-gesture agent) in her/his speech (e.g., [14, 21]). It was hypothesized that these gestures might have the property of "disambiguating" and clarifying the content of speech (e.g., [22, 23]). Butterworth and Hadar [24] argue that beats (small single movements stressing the words, as described in [25]) may improve communicative effectiveness of emphatic speech.

Speaker's gesturing can elicit some kind of meaningful reaction on the receiver, i.e., on his/perception of the speaker's and the message's characteristics, such as its effectiveness and persuasiveness. In a previous study ([16], by manipulating only gestures (through five conditions: ideational, conversational, object-, self-adaptors, and absence of gestures) of a female speaker in a video-message, it was shown that gesturing deeply affect the perceivers' explicit evaluation of both the message and the speaker: these effects concern, for instance, the perception of message persuasiveness, as well as the speaker's composure and competence and her communication style efficacy. These results were based only on data obtained through self-administered questionnaires after presentation of pictures and/or videos, and were providing information only on explicit measures, which are meant to capture the more cognitive and evaluative aspects within the general attitude [26] and might be not fully concordant with the "implicit" one. One possible way to get hints on the implicit attitude is by measuring the changes of those physiological variables which have been consistently proposed as correlates of positive or negative attitude towards stimuli with a different degree of pleasantness.

The most consistent data concern the activity of two facial muscles, the zygomatic major and the corrugator supercilii, respectively shown to be correlated to the perception of pleasant and unpleasant stimuli [27, 28]. Data had been also collected about changes in the activity of the two abovementioned muscles as a function of individuals' reaction towards verbal messages with positive or negative content (see, for example, [29, 30]).

A major issue raised by this body of evidence is whether these effects depend, partly or totally, on changes of vigilance and attentional levels. Since, to our knowledge, no study has yet investigated this aspect, it is necessary to think about possible measures of arousal and vigilance/attention complementing the assessment of muscular activity.

To these purposes, heart rate represents a widely used variable for the measurement of emotional arousal; it has been shown that changes in heart rate parallel the vision of emotionally involving movies, either pleasant or unpleasant [31], as well as the involvement in stress-inducing tasks such as public speaking [32].

In addition, the relationship between the perception of different kinds of gestures and spontaneous eye blinking may also be taken into account. An increase of spontaneous eye blinking has been consistently found for complex attention-demanding situations where executive functions are involved, such as elaborating visual stimuli [33] or being engaged in a conversation [34]. Pivik and Dykman [35] actually propose that blinks are regulated in conjunction with selective attention processes. Furthermore, the information-related aspects of stimuli might be involved in triggering spontaneous blinks [36]. Thus, by looking at eye blinks, it can be better understood whether any effect of gesturing is mainly conveyed via modifications of attentional system engagement or, alternatively, whether it depends more on a pure emotional channel, less modulated by cognitive higher functions.

The present study aims to investigate: (a) how gestures accompanying a disagreeable verbal message affect receivers' emotional reactions; (b) how strong and weak supporting arguments affect the intensity of the reactions and (c) how and to what extent these reactions can be modulated by different kind of gestures accompanying the verbal content. Access to the emotional attitude was provided by the measurement of four physiological variables: heart rate, eye blink rate, zygomatic major activity and corrugator muscle activity. We are especially interested in this latter measure, because it is a reliable index of the intensity of negative reactions [30].

The working hypotheses were that:

(1) A disagreeable content evokes a negative emotional reaction (i.e., it increases corrugator muscles activity);
(2) The intensity of a negative reaction tends to increase during the message, in particular in the middle of the speech;
(3) The negative reactions to the message are modulated by different gestures and different (strong vs. weak) arguments in support of a disagreeable topic;
(4) Such reactions are modulated by different gestures, which evoke different type of feelings (positive and negative), on their turn indexed by the muscular activity of two facial muscles.

The effects of different gestures on each of the abovementioned physiological variables were tested. It is expected that ideational gestures would increase the intensity of negative reactions to the message, because, being tightly related to speech semantics, they would enhance the impact of the unpleasant content. Instead, other gestures, even though affecting physiological arousal and/or attention, would not significantly change receiver's attitude, being capable to divert receivers' attention from the message's verbal content either to its form and its prosodic aspects (conversational gestures) or the speaker's body actions (adaptors).

8.2   Method

Participants. Fifty healthy subjects (35 females, 15 males), aged 19–37 (mean age 22,2 ± 2,8), all University students, participated in the study. They were randomly selected from a wider list of subjects who had given their availability and their informed consent prior to the inclusion in the study.

All subjects satisfied the following inclusion criteria: (a) no history of psychiatric illness; (b) normal physical status, without any major medical, neurological and/or ophthalmological illness; b) at time of testing, no cold, flu, headache or any clinical conditions impairing vigilance and attention; (c) no contact lenses or any other condition interfering with visual perception.

Materials. In this study the same source of stimuli was used as in a previous research of ours [16], namely a video message, in which a professional actress provides arguments supporting the introduction of a 20% increase of university fees for the Faculty attended by students participating to the experiment.

The increase of university fees was selected as experimental topic after running a preliminary survey, consisting of a series of qualitative interviews and questionnaires aimed to evaluate the students' degree of involvement, knowledge, and agreement about a wide number of university issues. It was found that the message's verbal content on the increase of university fees was perceived as relevant by the students and they showed a negative attitude towards it: in fact, students in the preliminary survey generally showed, towards this proposal (on a scale ranging from 1 to 7), medium-high interest (4.88 ± 1.99), a medium degree of knowledge (4.41 ± 1.94) and low agreement (2.04 ± 1.16).

The verbal message structure was defined by selecting four key points supporting the proposal (two strong and two weak arguments): therefore, from the student preliminary survey, we selected two strong points which had evoked medium-high agreement ($m > 4.85$) and high interest ($m > 5$), whereas other two (weak) points, not coming out from the qualitative interviews and absent in the preliminary survey, were created *ad hoc*.

Five experimental conditions were prepared, differing from each other only for the gestures acted by the actress during her speech, whereas the content of the message was kept identical. Four types of gestures were selected and classified, according to the most common taxonomical systems [3, 18, 19]:

(1) conversational gestures (cohesive and beats), linked to the speech structure and rhythm; (2) ideational gestures (i.e., illustrators and emblems), linked to the speech content; (3) object-adaptor gestures, that imply touching objects; (4) self-adaptor gestures, that imply touching parts of one's own body.

Hand gestures, whose number was the same in each gesture condition, were scripted ahead and occurred synchronously with the speech (see the Appendix in Maricchiolo et al. [16], for a detailed description).

In a fifth condition, the same speech was delivered in absence of gestures (control condition).

An accurate description of the stimuli and of specific gestures performed in each condition can be found in the Appendix of Maricchiolo et al. [16].

Measures. Subject's poly-graphic recordings were obtained by means of a Grass Model 78 polygraph, and included the following channels: (1) Vertical Electrooculogram (EOG); (2) Horizontal EOG. Both 1 and 2 were used to detect spontaneous eye blinks. To this aim, 0.25 cm gold skin electrodes filled with electrolyte paste were placed 3.0 cm above and 2.0 cm below the left eye, as measured from the center of the pupil to the center of the electrode (vertical EOG), and at the outer canthi (horizontal EOG). An eye blink was defined, according to previous studies [32–34], as a sharp high amplitude wave higher than $100 \mu V$ and no longer than 400 ms. Blink rate corresponds to the number of eye blinks/minute; (3) Mean electromyographic (EMG) activity of zygomatic major muscle (4) Mean electro-myographic (EMG) activity of corrugator supercilii muscle. Electromyograms were recorded by placing

0.25 cm gold skin electrodes respectively on the left cheek and at the median end of the left eyebrow, according to Fridlund and Cacioppo guidelines [37]. Raw EMG signals were amplified through a 10.000 gain, with high- and low-pass filter settings at respectively 50 and 500 Hz, and subsequently averaged in 1-s intervals; (5) Electrocardiogram (EKG), to measure heart rate, which was calculated by each R-R interval. Interbeat intervals were reduced off-line to heart rate in beats per minute.

Signals' digitalization of the signals was carried out through the Grass Polyview software according to conventional method (i.e., Codispoti et al. [28]).

Procedure: Ten subjects were randomly assigned to each of the five experimental conditions, making sure that the number of females and males would be the same in all conditions (F = 7, M = 3). All the experimental sessions were carried between 11 AM and 1 PM. Lighting conditions were held constant by excluding sunlight through black curtains and using artificial dim lights at the back of subjects' seat.

At arrival, subjects were prepared for polygraphic recordings and, after the electrodes' montage, instructed to sit as comfortably as they could in front of a flat monitor screen (Fig. 8.1).

Polygraphic recordings had a total duration of ten minutes, according to the following schedule: five minutes were allowed for habituation, then subjects were shown a 150 s naturalistic documentary, showing Arctic animals in a landscape without any people and requiring low arousal levels and low attentional engagement. This was used as baseline reference (BL video). Soon after the end of the documentary, subjects were shown the 150 s experimental video (EXP video). All biosignals were continuously recorded throughout the whole trial in "real-time" measurements, as illustrated in Fig. 8.1, and the abovementioned dependent variables were obtained by taking into account the central 120 s of each recording.

Design and Data Analysis. Four 2 × 5 ANOVAs (Neutral video—BL—vs. Experimental Video—EXP—x 5 gesture conditions) were carried out to test the effects of each category of gesture on each physiological variable. In order to look at the

Fig. 8.1 Experimental setting

progressive change of listeners' physiological measures while watching the videos, indexing modifications of arousal, attention and positive/negative feelings, we subdivided the 2-min recordings in 10-s chunks at regular intervals, obtaining 12 time-points along the speech. In such a way, we could carry out a 12×5 ANOVA (time points x gesture conditions).

To test the hypothesis regarding the effects of strong and weak arguments on negative reactions, we subdivided the central 120 s of the polygraphic recordings in five 20-s epochs, each corresponding to the presentation of a specific part of the experimental video: T1 topic introduction (*"First of all I'd like to express my opinion about an increase of University fees. I think there are many reasons to support the proposal of raising University fees by 20 %."*); T2 strong argument (*"It would improve University teaching and the didactic organization of theoretical units, laboratories, and exams. It could improve both the theoretical knowledge and the practical competences of prospective graduates and give them more work opportunities"*); T3 weak argument (*"It could push the development and progress of Italian scientific research in the psychological field. It will allow to keep pace with the other international scientific communities regarding both knowledge and results."*); T4 weak argument (*"thanks to this administrative action the Faculties of Psychology could become comparable to other scientific faculties such as Chemistry, Biology, Medicine and others, which have already considered a 20 % increase University fees"*); T5 strong argument (*"The increment in fees would permit the improvement of premises and a higher environmental quality. Rooms and facilities would be more adequate; laboratories could be better equipped, libraries would be richer; bulletin boards would be more easily accessible; secretarial offices and Internet information service would become more efficient"*). An ANOVA 5×5 (gesture conditions x arguments) was carried out on the corrugator activity to test the interaction effects of gestures with the different arguments presented during the message on the degree of negative reactions.

8.2.1 Results

1. Effect of gesture videos on physiological measures

As for those variables indexing arousal and attention, namely Heart Rate (beats/min) and eye blinking rate (blinks/min), the Type of Video*Condition 2×5 ANOVA shows a significant effect of Type of Video (the experimental –from now on EXP—vs. the baseline one—from now on BL -) but no significant Type of Video* Condition interaction (*Heart rate*: Type of Video $F1, 45 = 39.1$, $p < 0.0001$; Type of Video*Condition $F4, 45 = 0.7$, n.s.; *Eye Blink Rate*: Type of Video $F1, 45 = 14.7$, $p < 0.001$; Type of Video*Condition $F4, 45 = 1.2$, n.s.). As shown in Table 8.1, the mean values ($\pm$S.D.) of heart rate (indexing arousal) and eye blink rate significantly increased in EXP relative to BL for all categories of gestures.

For the abovementioned measures, ANOVA again detects a significant effect of Type of Video, but no significant Type of Video*Condition interaction, on the EMG activity of the corrugator muscle, indexing negative feelings (Corrugator Supercilii:

Table 8.1 Differences between the EXP and BL videos in the heart rate (beats/min)—, index of arousal—and eye blinking rate (blinks/min)—index of attention—

Gestures' categories	Heart rate (beats/min, mean + SD)		Eye blink rate (blinks/min, mean + SD)	
	BL	EXP	BL	EXP
Control (no gestures)	80.0 ± 14.9	83.1± 13.1	19.1 ± 11.6	27.6 ± 15.6
Self-adaptor	77.0 ± 10.1	82.0 ± 14.0	19.6 ± 13.7	21.9 ± 11.0
Object-adaptor	78.5 ± 12.3	81.4 ± 12.8	17.8 ± 10.7	19.9 ± 11.6
Conversational	74.9 ± 13.5	79.2 ± 14.4	18.7 ± 9.6	21.9 ± 13.0
Ideational	82.5 ± 12.5	88.0 ± 8.6	15.5 ± 7.8	19.9 ± 6.3

Table 8.2 Differences between the EXP and BL videos in muscular activity of the zygomatic major muscle—index of positive feelings—and the corrugator supercilii muscle—index of negative feelings

Gestures' categories	Zygomatic major (mV, mean + SD)		Corrugator supercilii (mV, mean + SD)	
	BL	EXP	BL	EXP
Control (no gestures)	12.7 ± 4.0	14.0 ± 6.1	9.9 ± 3.5	12.8 ± 7.2
Self-adaptor	13.8 ± 5.2	14.0 ± 6.7	10.1 ± 6.7	11.6 ± 3.4
Object-adaptor	12.9 ± 3.9	14.1 ± 5.0	8.0 ±1.6	11.5 ± 4.4
Conversational	15.6 ± 7.0	15.2 ± 6.9	12.2 ± 7.9	13.3 ± 9.2
Ideational	14.2 ± 4.6	17.6 ± 9.2	9.8 ± 5.3	12.9 ± 13.3

Type of Video—$F_{1, 45} = 6.1$, $p = 0.02$; Type of Video*Condition—$F_{4, 45} = 0.2$, n.s.), and no significant effects either of Type of Video or Type of Video*Condition on the EMG activity of the zygomatic muscle, indexing positive feelings (Zygomatic Major: Type of Video $F_{1, 45} = 2.2$, n.s.; Type of Video*Condition $F_{4, 45} = 0.8$, n.s.). Therefore, the vision of the experimental videos is associated to an increased activity of the corrugator muscle only. Table 8.2 displays the mean values (±S.D.) of the zygomatic and corrugator supercilii muscle activity (measures are both expressed in mV) averaged over the total duration of the videos, in the EXP compared with the BL videos.

8.3 Time Course of Physiological Variables in Real-Time Measurement

The Condition*Time 5×12 Mixed ANOVAs showed that all variables were significantly affected by Time (Heart Rate: $F_{11,495} = 11.58$, $p < 0.0001$; Eye-Blink: $F_{11,495} = 25.47$, $p < 0.0001$; Zygomatic: $F_{11,495} = 2.642$, $p = 0.011$; Corrugator: $F_{11,495} = 3.58$, $p = 0.007$); however, only for the corrugator activity ($F(44, 495) = 3.410$, $p = 0.000$) there was a significant interaction effect (gesture x time point).

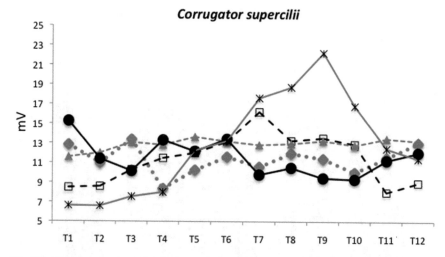

Fig. 8.2 Time-course of negative feelings for each experimental condition, obtained by measuring the corrugator muscle activity at twelve regular 10-s intervals (T1-T12). Legend: *rhombuses* self-adaptors, *squares* object-adaptors, *triangles* ideational, *dots* conversational, *crosses* no gestures)

The time-course of corrugator muscle activity, indexing implicit negative attitude while watching the video, is displayed in Fig. 8.2 for each experimental condition. This activity increases up to about the mid-point of the speech in the object-adaptor, ideational and control conditions; after that, it keeps stable at a plateau for ideational gestures, whereas it goes back to starting values in the no-gesture condition and decreases to intermediate values between the start and the mid-point peak for object-adaptor gestures. Despite the similar trends for the three conditions, an higher peak of negative emotions is reached during the no gesture speech.

8.4 Change of Physiological Measures in Response to Gestures and Arguments' Different Strength

A significant "gesture x arguments" interaction was found ($F(16,180) = 4.005$; $p < 0.001$; partial $\eta2 = 0.263$) on the corrugator muscle activity. Gestures interact with strong or weak arguments contained in the spoken message in determining negative reactions to the disagreeable message. In other terms, implicit listeners' reactions to strong or weak arguments are differently modulated according to what gestures categories are used by the speaker.

As displayed in Fig. 8.3:

- in the self-adaptors condition, no differences of corrugator activity appears in response to the different segments of the speech, corresponding to the introduction and the exposure of weak and strong arguments;

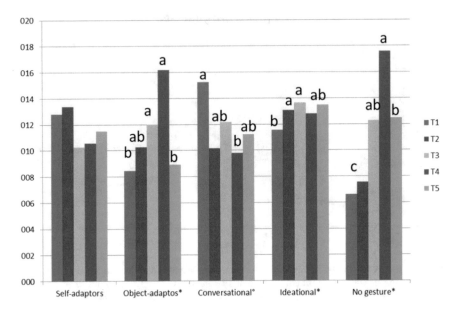

Fig. 8.3 Corrugator activity values in response to different segments of the speech, corresponding to the introduction and the exposure of weak and strong arguments in different gesture conditions (*T1* Introduction, *T2* Strong argument n.1, *T3* Weak Argument n., *T4* Weak Argument n.2, *T5* Strong Argument n.2). On y axis, corrugator muscle tone measured in millivolts)

- during the no gesture condition and the object-adaptors condition, receivers had a higher degree of the corrugator activity after weak arguments, especially at T4, rather than after strong ones and topic introduction.
- In the conversational gestures condition, receivers tend to increase the corrugator activity, indicating a negative reaction, during the introduction of the topic. The corrugator activity decreases while arguments in favor of an increase of the University fees are presented. This is particularly true for the strong ones.
- In the ideational gestures condition, the corrugator activity is rather flat with no substantial differences among the arguments.

8.5 Discussion

A significant increase at the experimental gesture video (EXP video), relative to baseline (BL video), was found in the levels of arousal, attention and negative thinking/emotions, as indexed respectively by an increase of the heart rate, the eye blink rate and the corrugator muscle activity respectively. These results are in line with our expectations, since the verbal content of the spoken message was a sequence of political arguments in favour of university taxes, representing a negative news for the university students to whom the message was submitted. Together with the absence of any significant increase at the experimental video in positive feelings—assessed through the zygomatic muscle activity—this evidence show that the proposed negative message had overall a strong negative impact on the receivers, as we had planned during the experimental design.

Indeed, we were particularly interested on listeners' negative reactions. To this regard, a very intriguing picture emerges from our analysis on the time profile of the corrugator muscle activity, based on subdividing the recordings durations in 10-s chunks, in order to look at the unfolding over time of the physiological measures (and therefore of the implicit psychological variables indexed by them) while subjects are receiving the message. Over the course of the video message, negative feelings and thinking are differently modulated by gestures. In fact, the verbal message increased negative emotions and thinking in the "no gestures" condition, with a very high peak reached at three quarters of the speech (T9) and a successive decrease, possibly due to habituation, which does not bring the corrugator muscle activity to the initial levels. This is not really surprising, since it is a control condition with the following negative characteristics: (a) no gesture may interfere with the direct effect of the verbal content; (b) a very low degree of ecological validity: at an impressionistic level, the actress who is not producing any gesture during her talk is probably perceived as highly unnatural. These two elements would then potentiate each other and cause the steep increase of the corrugator activity, up to a moment when a certain degree of habituation is reached to the unusual message's style. This habituation slightly reduces the negative attitude toward the spoken message, even though it remains higher than before the message was presented. We believe that this is because the absence of gestures makes the message less persuasive and the speaker less competent than when gestures are exploited, as also previously shown by a previous study of ours on explicit measures [16].

The significant "time*condition" interaction is mainly due to the increase of the corrugator muscle activity in the "object-adaptor" condition: after a progressive rise up to the speech mid-point, such activity returns to initial levels. Instead, the corrugator activity during ideational gestures is rather flat. It is possible to speculate on the mechanism beneath this particular pattern by taking into account what is known on the effects of object-adaptor gestures on explicit judgments: in fact, a female speaker producing object-adaptor gestures is evaluated quite positively as far as her competence, style effectiveness and persuasiveness are concerned, but at the same time she is judged rather negatively with respect to her composure, i.e., she is seen insecure, nervous and uncomfortable [16]. Therefore, the following sort of "two-steps" pattern can be hypothesized: in the beginning, object-adaptor gestures enhance the perceiver's unpleasant feelings, such as fear, danger, suspiciousness and displeasure, because they convey an impression of discomfort; as the speech goes by, this impression is progressively replaced by an opposite one of competence. Also, it cannot be ruled out that the initial enhancement of the negative attitude in the object-adaptor condition depends, in line with our initial working hypotheses, on an attentional shift from the verbal content to the formal features of gesturing.

As for ideational gestures, they have strong connections with the verbal message and can be regarded as "semantic gestures" (see [3], among others) because they repeat the information provided with the speech. Therefore, they contribute to clearly illustrate the content of the talk, which ends up in being more persuasive along the unfolding of the talk: in other words, despite the negative content is correctly elaborated and affects the overall level of the corrugator activity, which is higher than at the baseline video, there is no further increase over time because the speaker is perceived as persuasive and competent.

The overall scenario becomes more complex and intriguing when looking at conversational gestures which show, also at the time course analysis and at the "gesture * argument" interaction analysis, a remarkable "counterbalancing" effect on the negative content of the verbal message, differently from other types of gestures. It may be worthwhile noting that, opposite to the ideational, conversational gestures have no semantic link with the speech content, since they are mainly connected to the formal, non-propositional, aspects of the message [3].

Finally, analyzing how different gestures interact with the different arguments of the spoken message, we found that there is a modulating effect of them on the negative impact of the different arguments on the receivers. Weak arguments negatively affect receivers in the object-adaptors condition. Instead, in the conversational gesture condition, negative receivers' reactions tend to increase during the topic introduction, while decrease during the arguments' presentation, in particular the strong ones. Therefore, it can be hypothesized that the aforementioned counteracting effect of conversational gestures is specifically concentrated in the more meaningful parts of the message (the strong arguments being the more unpleasant and consequently the more likely to be contrasted by non-semantic gestures). Furthermore, in the absence of gestures, listeners' negative reactions increase along the whole speech, and in particular during the weak arguments' presentation.

8.6 Conclusions

This study represents a first experimental attempting to investigate negative emotional implicit reactions to a disagreeable verbal message by means of physiological indicators, taking into account: (a) the global role of gestures on these reactions; (b) how gestures and reactions interact over time during the message reception; (c) how gestures modulate the presentation of different kind of arguments supporting the message.

Our results demonstrates that receivers were globally aroused and attentive—as indexed by increased heart rate and eye blink rate, respectively, in the experimental videos compared to the baseline video—but negatively affected by the spoken message, with respect to the baseline videos (proved by the increase of EXP vs. BL. on corrugator activity and by the absence of any increase in zygomatic muscle activity). However, some gestures categories, in particular the conversational ones, seem able to partly reduce the negative impact of the verbal content, as shown by the time course of the corrugator muscle activity during the message presentation. Such results are of utmost importance when negative messages must be communicated, for example in crisis communication, designed to protect and defend an individual, company or organization facing a public challenge to its reputation, when an unpredictable event threatens important expectancies of other persons and can seriously impact an organization's or individual's performance and generate negative outcomes [38, 39]. Other contexts of application can be those situations in which unpleasant information (e.g., medical test results, layoff, etc.) must be reported.

The present panel of results appears relevant to the development of intelligent algorithms for human-machine interaction that are requested to sense humans not only to correctly satisfy their requests but also to mediate their perception of the whole interaction. In this context, gestures may play a significant role to make these systems "socially believable".

References

1. Hadar U, Butterworth B (1997) Iconic gestures, imagery and word retrieval in speech. Semiotica 115:147–172
2. Kendon A (2004) Gesture: visible action as utterance. Cambridge University Press, Cambridge
3. McNeill D (1992) Hand and mind. The University of Chicago Press, Chicago
4. Goldin-Meadow S, Alibali MW, Church RB (1993) Transitions in concept acquisition: using the hand to read the mind. Psychol Rev 100:279–297
5. Bavelas JB, Gerwing J, Sutton C, Prevost D (2008) Gesturing on the telephone: independent effects of dialogue and visibility. J Mem Lang 58:495–520
6. Graham JA, Argyle MA (1975) Cross-cultural study of the communication of extra-verbal meaning by gesture. Int J Psychol 10:57–67
7. Holler J, Shovelton H, Beattie G (2009) Do iconic gestures really contribute to the semantic information communicated in face-to-face interaction? J Nonverbal Behav 33:73–88
8. Kelly SD, Church RB (1998) A comparison between children's and adults' ability to detect children's representational gestures. Child Dev 69:85–93
9. Kendon A (1995) Gestures as illocutionary and discourse structure markers in Southern Italian conversation. J Pragmat 23:247–279
10. Goodrich W, Hudson Kam CL (2009) Co-speech gesture as input in verb learning. Dev Sci 12(1):81–87
11. Dick AS, Goldin-Meadow S, Hasson U, Skipper JI, Small SL (2009) Co-speech gestures influence neural activity in brain regions associated with processing semantic information. Hum Brain Mapp 30(11):3509–3526
12. Maricchiolo F, De Dominicis S, Ganucci Cancellieri U, Di Conza A, Gnisci A, Bonaiuto M (2014) Co-speech gestures: structures and functions. In: Müller C, Cienki A, Fricke E, Ladewig SH, McNeill D, Bressem J (eds) Body - language - communication. An international handbook on multimodality in human interaction, vol 2. De Gruyter Mouton, Berlin, pp 1461–1473
13. Beattie G, Shovelton H (2001) How gesture viewpoint influences what information decoders receive from iconic gestures. In: Cave C, Guaitella I, Santi S (eds) Oralite et Gestualite: Interactions et comportements multimodaux dans la communication. L'Harmattan, Paris, pp 283–287
14. Beattie G, Shovelton H (2006) When size really matters: how a single semantic feature is represented in the speech and gesture modalities. Gesture 6(1):63–84
15. Kelly SD, Ward S, Creigh P, Bartolotti J (2007) An intentional stance modulates the integration of gesture and speech during comprehension. Brain Lang 101:222–233

16. Maricchiolo F, Gnisci A, Bonaiuto M, Ficca G (2009) Effects of different types of hand gestures in persuasive speech on receivers' evaluations. Lang Cogn Process 24:239–266
17. McClave E (1994) Gestural beats: the rhythm hypothesis. J Psycholinguist Res 23:45–66
18. Ekman P, Friesen WV (1969) The repertoire of nonverbal behavior. Semiotica 1:49–98
19. Maricchiolo F, Gnisci A, Bonaiuto M (2012) Coding hand gestures: a reliable taxonomy and a multi-media support. In: Esposito A, Esposito AM, Vinciarelli A, Hoffman R, Müller VC (eds) Cognitive behavioural systems, vol 7403. Springer, Berlin, pp 405–416
20. Berger KW, Popelka GR (1971) Extra-facial gestures in relation to speech reading. J Commun Disord 3:302–308
21. Riseborough MG (1981) Physiographic gestures as decoding facilitators: three experiments exploring a neglected facet of communication. J Nonverbal Behav 5:172–183
22. Beattie G, Shovelton H (2005) Why the spontaneous images created by the hands during talk can help make TV advertisements more effective. Br J Psychol 96:21–37
23. Holle H, Gunter TC (2007) The role of iconic gestures in speech disambiguation: ERP Evidence. J Cogn Neurosci 19:1175–1192
24. Butterworth B, Hadar U (1989) Gesture, speech and computational stage: a reply to McNeill. Psychol Rev 96:168–174
25. McNeill D (1985) So you think gestures are nonverbal? Psychol Rev 92:350–371
26. Greenwald AG (1990) What cognitive representations underlie social attitudes? Bulletin of the psychonomic society 28:254–260
27. Blascovich J (2000) Using physiological indexes of psychological processes in social psychological research. In: Reis HT, Charles M (eds) Handbook of research methods in social and personality psychology. Cambridge University Press, New York, pp 117–137
28. Codispoti M, Bradley MM, Lang PJ (2001) Affective reactions to briefly presented pictures. Psychophysiology 38:474–478
29. Dimberg U, Thunberg M, Elmehed K (2000) Unconscious facial reactions to emotional facial expressions. Psychol Sci 11:86–89
30. Cacioppo JT, Petty RE (1981) Electromyograms as measures of extent and affectivity of information processing. Am Psychol 36:441–456
31. Gomez P, Zimmermann P, Guttormsen-Schar S, Danuser B (2005) Respiratory responses associated with affective processing of film stimuli. Biol Psychol 68:223–235
32. Al'Absi M, Bongard S, Buchanan T, Pincomb GA, Licinio J, Lovallo WR (1997) Cardiovascular and neuroendocrine adjustment to public speaking and mental arithmetic stressors. Psychophysiology 34(3):266–275
33. Fukuda K, Stern JA, Brown TB, Russo MB (2005) Cognition, blinks, eye movements, and pupillary movements during performance of a running memory task. Aviat Space Environ Med 76 (Suppl 7):C75–C85
34. Al-Abdelmunem M, Briggs ST (1999) Spontaneous blink rate of a normal population sample. Int Contact Lens Clinic 26:29–32
35. Pivik RT, Dykman RA (2004) Endogenous eye blinks in preadolescents: Relationship to information processing and performance. Biol Psychol 66:191–219
36. Fogarty C, Stern JA (1989) Eye movements and blinks: their relationship to higher cognitive processes. Int J Psychophysiol 8(1):35–42
37. Fridlund AJ, Cacioppo JT (1986) Guidelines for human electromyographic research. Psychophysiol 23:567–589
38. Coombs WT (2007) Ongoing crisis communication: planning, managing and responding. Sage, Los Angeles
39. Gonzales-Herrero A, Smith S (2008) Crisis communication management on the web: how internet-based technologies are changing the way public relations professionals handle business crisis. J Conting crisis Manage 16(3):143–153

Chapter 9
Laughter Research: A Review of the ILHAIRE Project

Stéphane Dupont, Hüseyin Çakmak, Will Curran, Thierry Dutoit, Jennifer Hofmann, Gary McKeown, Olivier Pietquin, Tracey Platt, Willibald Ruch and Jérôme Urbain

9.1 Introduction

Laughter is everywhere. So much so that we often do not even notice it. First, laughter has a strong connection with humour. Most of us seek out laughter and people who make us laugh, and it is what we do when we gather together as groups relaxing and having a good time. But laughter also plays an important role in making sure we

S. Dupont (✉) · H. Çakmak · T. Dutoit · J. Urbain
University of Mons, Mons, Belgium
e-mail: stephane.dupont@umons.ac.be

H. Çakmak
e-mail: huseyin.cakmak@umons.ac.be

T. Dutoit
e-mail: thierry.dutoit@umons.ac.be

J. Urbain
e-mail: jerome.p.urbain@gmail.com

W. Curran · G. McKeown
Queen's University, Belfast, UK
e-mail: w.curran@qub.ac.uk

G. McKeown
e-mail: g.mckeown@qub.ac.uk

J. Hofmann · T. Platt · W. Ruch
University of Zurich, Zurich, Switzerland
e-mail: j.hofmann@psychologie.uzh.ch

T. Platt
e-mail: tracey.platt@psychologie.uzh.ch

W. Ruch
e-mail: w.ruch@psychologie.uzh.ch

O. Pietquin
University of Lille, Villeneuve d'Ascq, France
e-mail: olivier.pietquin@univ-lille1.fr

© Springer International Publishing Switzerland 2016
A. Esposito and L.C. Jain (eds.), *Toward Robotic Socially Believable Behaving Systems - Volume I*, Intelligent Systems Reference Library 105,
DOI 10.1007/978-3-319-31056-5_9

interact with each other smoothly. It provides social bonding signals that allow our conversations to flow seamlessly between topics; to help us repair conversations that are breaking down; and to end our conversations on a positive note.

Currently, attempts are being made to create computer agents that interact with humans in the same way humans do, understanding their social signals (cf. review by Vinciarelli et al. [133]). However, laughter has not received as much attention in the area of Human Computer Interaction (HCI), despite its huge importance in social interactions, and as we will see, being one of the most important non-verbal vocal social signals. Gathering new knowledge on laughter, from both engineering and human sciences perspectives, and transferring this understanding to the design of computer agents will enable them with both emotional and conversational competence that will increase their impact and application potential. This was the goal of the ILHAIRE[1] collaborative research project. It has been supported by the Future and Emerging Technologies branch of the 7th framework program for research of the European Union. It started in September 2011 and ran until September 2014. It included an interdisciplinary team in different areas of human sciences and engineering: psychology of laughter, emotion-oriented computing, automatic recognition and synthesis of communicative behaviors and signals, study of non-verbal social communication cues, and virtual agents.

This chapter provides a summarized account of the context of this research topic and a comprehensive overview of the ILHAIRE research results, as well as references to pioneering works. It is organized as follows. Section 9.2 covers the roles and functions of laughter, during both hilarious as well as conversational interactions. Such knowledge is fundamental when reaching towards endowing computer agents with the capability to understand and apply this social signal. Section 9.3 covers another fundamental building block, the more precise understanding of the different characteristics of the signal itself, and how they affect the way laughter is perceived and impacts us, together with the personality factors that matter. Such studies are heavily relying on observations drawn from multimodal corpora of human-human and human-computer interactions. The advances in available databases and their annotation are reported in Sect. 9.4. Corpora are also key in designing computational approaches enabling to embed laughter in HCI, which imply methods to accurately detect and recognize laughter, to produce natural-looking and natural-sounding laughter, to perceive the communication scene and make use of the communication context in order to laugh the right way at the right time. These are respectively addressed in Sects. 9.5, 9.6 and 9.7. They cover the different facets of laughter, in particular the sound, the gestures and body motion/posture and the facial expressions. Finally, application perspectives are drawn in Sect. 9.8 while research perspectives are outlined in Sect. 9.9.

---

[1] http://www.ilhaire.eu/.

9.2   Roles and Functions

Laughter is an ubiquitous social signal in human interaction; it occurs very often and only a fraction of the occurrences seem to be related to humour. The ILHAIRE project made an initial distinction between laughter that was social and conversational in nature and laughter that was hilarious, with the latter kind of laughter being more directly related to humour. The research conducted within ILHAIRE bore out this distinction, with laughter intensity being one important distinguishing variable. To understand why these separate forms of laughter may have arisen, and what functions each serves, we can look to their evolutionary origins and to the situations in which social and hilarious laughter are found.

Laughter is an evolutionarily ancient behaviour which arguably precedes spoken language [18, 19]. This is borne out by the observation of laughter-like behaviours in many primate species. Preuschoft and van Hooff [101] argue that the smile and laughter distinction has its origin in two different sets of signals, and that both sets of signals have important social functions in regulating and enhancing social affiliation.

Laughter-like behaviour in chimpanzees prolongs play actions [70], suggesting that it is an important tool for promoting social affiliation and developing cooperative and competitive behaviours [19], and similar behaviour is observed in human infants [106].

Smile-like behaviours have been observed in primates signalling affiliation, appeasement, reassurance, and submission; again a social bonding element seems to unify these functions. Thus from an evolutionary perspective, laughter and smiling can be viewed as key adaptive behaviours because of their facilitative effect on social cohesion. Preuschoft and van Hooff [101] propose that, within a number of primate species and especially humans, the boundaries between smile- and laughter-like signals have become blurred and intensity-related phenomena enable an ambiguous yet gradual transition of function from social and (in the case of humans) conversational laughter to stronger amusement related laughter.

Social and conversational laughter predominates at low levels of intensity. Here there seems to be a strongly ordered and rule bound nature to laughter production. These rule sets have been most firmly elucidated within the Conversation Analysis tradition [36, 50, 88], but also confirmed in some more experimental work [7]. Laughter in conversation tends to be initiated by one member of an interaction, but others can join in and it is typically a shared rather than competitive activity. Laughter can be received as an invitation to shared laughter that can lead to reciprocated laughter, or responded to with silence or serious talking. A response to silence can be the further pursuit of shared laughter while serious talking is an active declining of the invitation [36]. While the latter response can often extend the exploration of a topic of conversation, accepting an invitation to laughter can lead to a topic change. In this way laughter has a regulatory function in conversations by serving as a turn-taking cue or signaling that the speaker may be approaching a transition point in their topic or theme [50, 88]. Other conversational rules suggest that in dyadic conversations the speaker is more likely than the listener to laugh first, while in group conversations the listeners are more likely to laugh first [36].

An important feature of laughter that facilitates these conversational dynamics is that it has an inherent ambiguity. This is most apparent at low levels of intensity, but the transition to higher levels of intensity (related to exhilaration) also seems to be reasonably weakly defined. Within the ILHAIRE project experimentation showed that, when removed from its conversational context, laughter becomes hard to classify in terms of classic emotional or sociolinguistic labels. This suggests that context plays a strong role in the interpretation of laughter, as has been recently highlighted in more general valance and emotion-based stimuli [54]. McKeown et al. (under review) have argued based on ILHAIRE research that ambiguity plays an important role in servicing and maintaining social interaction. Holt [50] suggests that, as laughter as a signal has no propositional content, it can be safely interspersed within conversations; because of its lack of propositional content, laughter serves as a social bonding signal that allows an abstraction from the content of the conversation, it is this that allows safe turn-taking transitions, topic changes and terminations. McKeown et al. suggest that this ambiguous property of laughter allows multiple interpretations of the same content to be held by different interlocutors, which facilitates the repair of conversations. Further, they propose that laughter ambiguity also facilitates the holding of two separate interpretations of the same content in the minds of both interlocutors at the same time. Consequently laughter can aid the safe exploration of possible taboo areas, hypothetical scenarios, and impropriety while retaining the possibility of plausible deniability. Laughter requires an ambiguous nature to enable this social exploration, allowing it to be abstracted from the content and to be easily interpreted in multiple ways. The ambiguous nature of laughter has been underlined by an important study within the ILHAIRE project, in which similar-intensity laughter from different contexts within the same conversation, as well as from different conversations, was interchangeable with minimal impact on the perceived genuineness of the conversational interaction.

Laughter also has clear relationships with humour. In ratings of laughs extracted from their conversational context, consistent strong positive correlations (in the range of $r = 0.65$ to $r = 0.7$) were found between ratings of intensity and rating of association with humour. Not surprisingly, high intensity laughter appears to be strongly related to humour. McKeown [76], has argued that humour production is a hard-to-fake signal of creativity (following Miller [79] and Greengross and Miller [38]) and of mind-reading ability. Correspondingly, McKeown et al. [74, 75] have argued that laughter also serves as a hard-to-fake signal of humour appreciation. The construct of exhilaration indeed describes the effective response to humour [107, 110]. The laugh features that arise with increasing laugh intensity are important to convince an interlocutor that the laughter is related to a genuine felt emotion state. Yet, the boundaries of where this transition to laughter associated with the felt state of amusement and laughter that serves more socio-communicative functions remain ambiguous.

Beside felt amusement following humour, laughter is actually related to other enjoyable emotion states. Ekman [28] proposed that there are at least sixteen enjoyable emotions rather than simply a global emotion of joy. He hypothesized that even though they would all go along with the Duchenne display, namely, the joint and symmetric contraction of the zygomatic major and orbicularis oculi muscles (pulling the lip corners back- and upwards and raising the cheeks and compression of the eyelids causing eye wrinkles, respectively), the main differences in the facets of joy would lay in the parameters, such as timing of onset and offset or intensity. While investigating the responses of those with a fear of being laughed at within ILHAIRE, Platt et al. [100] discovered that, of those sixteen emotions, some were consistently associated with the expression of laughter (e.g., amusement, relief, tactile pleasure, schadenfreude).

Finally, although there are individual differences in the susceptibility or willingness to engage in the moment, signals that are displayed when positive emotions are being experienced are contagious, in as much as others who decode those signals will often feel enjoyment [41]. Again, laughter has a strong role in maintaining social connection.

Overall, the roles and functions of both conversational and enjoyment laughter, and the laughter linked to the experience of and the contagion of positive emotions will be key for the smooth interaction with virtual agents.

9.3 Characteristics, Perception and Effect

Laughter hence appears as one of the most important non-verbal vocal social signal. But beyond understanding its roles and functions and knowing the contexts where it is relevant, it will also be crucial to understand more precisely what are its different characteristics (also referred to as features here under), and how these affect the way the signal is perceived, and impact on people. Laughter can indeed be very varied, and although in many cases it has a strong inherent ambiguity, we will see that specific laughter features can have a significant impact on its perceived naturalness, emotional color (valence, arousal, dominance), maliciousness (or alternatively friendliness), and contagiousness. Besides, inter-individual difference in experiencing laughter signals have been identified, including people with a fear of being laughed at, as well people with autism spectrum disorders. It is hence fundamental when creating virtual agent equipped with laughter to understand which of the social signal facets need to be designed with care.

Works on laughter perception can be divided into subtopics, guided by the modality that was the focus of investigation (face, voice, body). Before ILHAIRE, most work has been done on the perception of only auditorily presented laughs. Within ILHAIRE, we have worked on all three modalities and investigated the perception of naturally occurring laughs (spontaneous), acted laughs, manipulated laughs, and virtually portrayed/synthesized laughs. Novel insights have been obtained, and some of those then guided technological and experimental developments, exposed in Sects. 9.5, 9.6 and 9.7.

9.3.1 Perception of Facial Features

Coeval writers of Charles Darwin, stemming from the historic German field of Ausdruckspsychologie (expression psychology; for example Piderit 1867 or Bore 1899, cf. [111]) delivered extensive descriptions of the vocal and facial markers of qualitatively different laughter types. Thus, the perception-related studies within ILHAIRE started with two investigations of the facial features of laughter basing on historic knowledge [49, 111]. The two studies concentrated on the facial features of four different laughter types (joyful, intense, schadenfreude laughter, and grinning). These four laughter types were chosen because (a) at least four historic authors had described them in their laughter classification, and (b) the authors had delivered a visual illustration as well as verbal description (cf. [111] for details). A total of 18 illustrations were first examined for their facial features with a technique allowing for the objective assessment of all visually discernible facial actions [29]. Then, the decoding of these laughter types by laypersons was investigated in two online studies. The results showed that illustrations of laughter involving a Duchenne Display (DD; the symmetric and simultaneous contraction of the zygomatic major muscle and orbicularis oculi, pars orbitalis muscle) were perceived as joyful laughter, irrespective of their initial classification by the historic writers. Only the DD configuration could be reliably morphologically differentiated and was recognized at high rates. In intense laughter, the intensity of the FACS coded zygomatic major muscle action predicted the perception of intensity by the laypeople. The proposed changes in the upper face highlighted in the literature, i.e., the presence of an additional eyebrow-lowering frowning, did not predict the perception of intensity. Even more, the presence of eyebrow-lowering frowning was antagonistic to the perception of joy. Schadenfreude and grinning did not have high recognition rates, but these displays were also highly heterogeneous in their portrayals. For schadenfreude laughter, two hypotheses were put forward [49]: Schadenfreude may either be a blend of a positive and negative emotion (entailing facial features of both), or expressed by a joy display with regulation or masking attempts (as it is not socially desirable to laugh at the misfortune of others, [28]). Hofmann and colleagues have tested these hypotheses in two encoding studies within the ILHAIRE project [45, 47], showing that indeed, schadenfreude was often dampened or down-regulated when expressed in social contexts. While many historic writers had claimed facial morphological differences from joyful to intense joyful laughter, our decoding studies did not support the proposed changes, but indicated that the presence of markers beyond the Duchenne markers did not increase the perceived intensity, but led to a change in the perceived valence of the laughter. If eyebrow-lowering frowning (a proposed marker of laughter intensity) was present, the laughter was consequently rated as more malicious. Therefore, we investigated this notion further within ILHAIRE by manipulating the presence or absence of eyebrow-lowering wrinkles in synthesized avatar laughter. Basing on synthesized laughter animations with refined facial wrinkles [81, 94], two studies were conducted to investigate the influence of the presentation mode (static, dynamic) and eyebrow-lowering frowning on the perception of laughter animations of different intensity [42]. In a first study, participants (N = 110) were randomly assigned to two presentation modes (static pictures versus dynamic videos) to watch animations of Duchenne laughter and laughter with added eyebrow-lowering frowning. Ratings on the intensity, valence, and contagiousness of the laughter were completed. In a second study, participants (N = 55) saw both animation types in both presentation modes sequentially. Our results confirmed that the static presentation mode lead to eyebrow-lowering frowning in intense laughter being perceived as more malicious, less intense, less benevolent, and less contagious compared to the dynamic presentation, just as we found in the study of historic illustrations. This was replicated for maliciousness in the second study, although participants could potentially infer the "frown" as a natural element of the laugh, as they had seen the video and the picture. Hofmann [42] concluded that a dynamic presentation is necessary for detecting graduating intensity markers in the joyfully laughing face. While these studies focused on general differences in laughter perception, we also investigated inter-individual differences within the ILHAIRE project. Hofmann and colleagues [44] studied the responses to photos of different smiles and laughter and found that gelotophobes assigned the joyfully laughing face not only joy, but also contempt. Thus, for gelotophobes, the "smiling face may hide an evil mind". Ruch and colleagues [113] looked at how the fear of being laughed at (gelotophobia, cf. [112] for a recent review on this topic) influenced the perception of laughter animations (face and upper body avatar portrayals, synthesized laughter sounds with four different modifications, faceless full body stick figure animations). For the perception of the face, the results showed that gelotophobes found medium intensity laughs that gave the impression of being contrived or regulated as most malicious. The shape and appearance of the lips curling induced feelings that the expression was malicious for non-gelotophobes and that the movement round the eyes, elicited the face to appear as friendly. This was opposite for individuals with a fear of being laughed at: they perceived those features as indicative of maliciousness.

9.3.2 Perception of Acoustic Features

Most previous studies on laughter acoustics have focused on the decoding of natural, posed, and manipulated laughs. These studies of laughter mainly followed the notion that single laughter elements[2] and changes in acoustic parameters are important for the identification and evaluation of a given laugh [57, 58]. In Table 9.1, findings on the perception of laughter features are summarized (adapted from [43]).

Table 9.1 shows that voicing is a potent predictor of the perception of the positive valence in laughter [2–4, 20]. Basing on this finding, many studies concentrated on voiced laughs and modified acoustic parameters of such laughs to investigate perceptual changes. F0 variations were found to influence the perception of valence, arousal, and dominance (see Table 9.1). For example, the descending pitch (F0) in successive laughter elements was evaluated as more friendly or genuine than laugh-

---

[2]Laughter elements correspond to individual bursts of energy, whose succession is characteristic of laughter.

Table 9.1 Findings on the perception of laughter acoustics

Dimension	Acoustic features	Stimuli
Naturalness/Realness/ Genuineness	• Serial patterns with varying parameters = ratings close to natural laughs	Natural + manipulated laughs [8, 56–58, 134]
	• Faster is perceived as more real/natural	
	• Series with stereotyped patterns are perceived as less natural and genuine	
	• Descending F0 in successive laughter elements is perceived more genuine than laughter series of elements with a constant pitch	
Arousal	• More rapid is perceived as higher aroused	Posed laughs [121]
	• Higher laugh rate is perceived as higher aroused	
	• Lower inter-bout duration is perceived as higher aroused	
	• Higher pitch is perceived as higher aroused	
	• Higher levels of high-frequency (HF) energy is perceived as higher aroused	
Dominance	• Higher intensity parameters more dominant	Posed laughs [121]
	• More precise articulation (lower F0 band width, lower jitter) more dominant	
	• Energy more strongly concentrated in the high frequency range more dominant	
	• Prolonged vocalic segments more dominant	
	• Temporal distance between bouts shorter perceived more dominant	
	• Lower harmonic energy (less voiced elements) perceived more dominant	
	• Dominance is predicted by the interval from vowel to vowel (58 %)	Natural + forced laughs [62]
	• Dominance is predicted by F0 (mean, max) of the noise vowel reiteration (31–31 %)	
	• Dominance is predicted by Small versus large amplitude diminishment (31 %)	
Valence	• Higher number of segments, higher laugh rate, lower inter-bout duration is perceived more positive	Posed laughs [121]
	• Voiced laughs are perceived more positive than unvoiced	Natural laughs, modified laughs [2–4, 20, 56–58]
	• Duration of the initial expiratory noise predicts 42 % of the positive valence rating	Natural, forced laughs [62]

F0 = pitch

ter series of elements with a constant F0 [56]. In subsequent work, [57] modified naturally occurring laughter in different laughter series. The results indicated that experimentally modified series with decreasing pitch and variable duration, as well as the series with sub-phrases, were evaluated as good as spontaneous laughter. In particular, series with decreasing parameter courses such as decreasing pitch or declining durations in successive elements were rated as "friendly" and "laugh-like". Also, series with rhythmic patterns "long-short" and "long-short-long" (accents within the rhythm of a laughter-series) evoked more smiles and laughs in listeners than all other series [56–58]. Moreover, stereotyped patterns in the course of the F0 received less good evaluations. Kipper and Todt [57] concluded that the evaluation of laughter depends on the dynamic changes of acoustic parameters in successive elements of laughter.

A different methodology was utilized by Tanaka and Campbell [122]. In their study, students first labeled laughter examples from natural conversations in four pre-defined laughter categories. Most laughs were categorized as sounding "polite" or "mirthful". Second, they performed an acoustic analysis of all the laughs that were labeled as "polite" or "mirthful" to identify acoustic features discriminating between these two types. Polite laughter was related to low maximum power and mirthful laughter to high maximum power, long duration, and a high number of bouts. They further reported that the best predictors of the two laughter types were the pitch (mean and maximum value), the number of bouts, power, spectral slope, and measures of prosodic activity. With these features, a classification was performed, leading to 79 % classification accuracy between polite and mirthful laughter. Thus, listeners distinguished between laughter types in auditory stimuli, which indicated that those types have a distinct signal value. Nevertheless, there were also group-related differences. For autistic individuals, differences in global laughter evaluations were found. Hudenko and Magenheimer [51] found that autistic childrens voiced laughs were perceived as more positive than normally developed childrens voiced laughs. The latter were generally lower pitched, and shorter (but no differences in F0 were found). When comparing individuals with a fear of being laughed at to individuals with no fear, Ruch and colleagues [113] found within ILHAIRE that the fundamental frequency modulations and the variation in intensity were indicative of perceived maliciousness. Fast, non-repetitive voiced vocalizations, variable and of short duration were perceived as most friendly by individuals with a fear of being laughed at.

9.3.3 Perception of Bodily Portrayals

Enjoyable emotions are aligned with laughter, that involves open mouth smiling, vocalized laughter sounds, but also bodily changes such as dropping or relaxing the shoulders, and shaking of the trunk. Within ILHAIRE, another strong focus did hence lay in the investigation of laughter in the body and the perception of such cues. Within ILHAIRE, Griffin and colleagues [40] analyzed participants perception of laughter from body movements. The participants task was to categories animations of natural laughter from motion capture data replayed using faceless stick figures (characters with trunk, limbs and heads simply represented by edges). In general, animations that were perceived as representing a laugh differed in torso and limb movements compared to stimuli categorized as non-laughter. Also, the distinguishing features differed for laughter stemming from sitting or standing avatar positions. Perceived amused laughter differed from perceived social laughter in the amount of bending of the spine. Similarly, Mancini et al. [67] found that laypersons were generally very good in distinguishing full-body animations of laughter from non-laughter (79.70 % of the stimuli were categorized correctly), with high levels of confidence in rating either stimulus category. When assessing the perceptions of gelotophobes and non-gelotophobes, Ruch and colleagues [113] found that in the virtual body portrayals (faceless full body stick figures), backwards and forward movements and rocking versus jerking movements distinguished the most malicious from the least malicious laugh.

9.3.4 Perception of Multimodal Portrayals

Sestito and colleagues [117] investigated the decoding of audio-visual laughter stimuli. They found that the correct decoding of laughter was high above chance rate and that in audio-visual incongruent stimuli, the visual modality was prioritized in the decoding over the acoustic dimension. Using electromyography measurements of the zygomatic major muscle activity, they also reported that rapid and congruent mimicry toward laughter stimuli occured. Within ILHAIRE, several studies have investigated the perception of multimodal portrayals of human laughter (spontaneous and acted/fake laughs; presented with visual-auditory stimuli). McKeown and colleagues [74] conducted two experiments to assess perceptions of genuine and acted male and female laughter and amusement facial expressions. The main results showed that participants were good in detecting fakeness in laughs by males. When women faked laughs, males distinguished cues of simulation, but judged fake laughs also to be more genuine. When judging other women, female participants perceived genuine laughs to contain higher levels of simulation. With a focus on inter-individual differences, Ruch et al. [114] investigated the verbal and facial responses of 20 gelotophobes and 20 non-gelotophobes towards videos of people recalling memories of laughter-eliciting positive emotions (amusement, relief, schadenfreude, tactile pleasure). The facial expressions of the participants were clandestinely filmed and evaluated by the FACS [29]. Smiles of enjoyment and "markers of contempt" were coded and verbal ratings of the participants obtained. Gelotophobes responded with less joyful smiles and with more expressions of contempt to laughter-eliciting emotions than did non-gelotophobes. Gelotophobes also perceived the degree of joy expressed by participants in the video clips of tactile pleasure and relief lower than non-gelotophobes. No differences occurred in the perception of joy for schadenfreude and amusement. Thus, spontaneous affective responses and cognitive responses through ratings have to be distinguished.

9.4 Naturalistic Databases

At the start of the ILHAIRE project, there were a small number of existing databases that provided examples of laughter for research purposes. The most useful of these were the AudioVisual Laughter Cycle database (AVLC) [127] and the MAHNOB laughter Database [97], both of which contain laughter from individuals watching funny video clips. While these databases serve their purpose well, they are limited to a certain style of laughter and context and both had similar aims in the style of laughter targeted. The ILHAIRE project sought to collect a much larger amount of laughter and from a much more diverse range of settings and contexts than had been previously gathered. The goals of the ILHAIRE laughter database were to collect laughter from a broad range of contexts. Thus in addition to collecting laughter of people observing amusing scenarios (e.g. watching comedy), we also targeted laughter occurring in social interaction and, importantly, in situations that led to what was termed hilarious laughter and what was deemed to be more social and conversational in style. There were further goals of collecting laughter data from more than one culture and from interactions that took place in more than one language. In addition the range of sensors that gathered the information was to be extensive, including high quality audio and video, but also incorporating depth, motion capture information, respiration and facial expression information where possible. The project also sought to provide as much annotation as resources would allow and devise an annotations scheme for laughter to facilitate this.

As a result of these diverse goals the ILHAIRE laughter database is not a database in the traditional sense but something of a meta-database; it incorporates a number of different databases and sources with the overarching goal of providing a useful laughter-focused set of resources and stimuli for the research community. There were three main phases to this: the collection and annotation of laugh stimuli from existing databases (also summarized in McKeown et al. [71]), the collection of hilarious laughter, and finally the collection of conversational and social laughter. The first phase is distinct due to the nature of the task. However, the other two phases are less distinct as it is not straightforward to define what distinguishes a hilarious laugh and a social laugh and these laugh types both commonly occur in any given social interaction and as argued on the basis of ILHAIRE research the boundaries between these laughs are ambiguous (underdetermined).

9.4.1 Existing Databases

Belfast Naturalistic Database

The Belfast Naturalistic Database [25] was an early attempt to gather a broad swath of audio-visual material of people who at least appeared to be experiencing genuine emotion with material sourced mainly from television programmes. 53 of the total of 127 video clips contain laughter, but only five can be made available due to copyright issues.

HUMAINE Database

The HUMAINE database [26] was created with the purpose of demonstrating the existing breadth of material related to a broad understanding of the word emotion—termed "pervasive emotion". From fifty video clips 46 instances of laughter of variable quality were extracted, and are useful as illustrations of the variety of situations in which laughter occurs.

Green Persuasive Database

The Green Persuasive Database [26] contains audiovisual clips recorded to capture interactions with strong feelings, but not basic emotions. The scenario involves one participant convincing another to adopt an environmentally friendly lifestyle. There is a strong power imbalance between participants as the persuader is a University Professor and the listeners are students. There were eight interactions in total lasting between 15 and 35 min. From these eight participants, 280 instances of laughter were extracted.

Belfast Induced Natural Emotion Database

The Belfast Induced Natural Emotion Database (BINED) [119] represents a deliberate effort to induce specific kinds of emotional behaviour. The goal was to use a series of tasks to generate spontaneous and dynamic emotional material that could replace the posed static photographs often used in studies of emotion. Laughs were extracted from Set 1 of the database including tasks designed to elicit: amusement, frustration, surprise, disgust and fear. 289 instances of laughter were extracted from a total of 565 clips with 113 participants (43 female, 70 male).

SEMAINE Database

The SEMAINE database [77] provides high quality audio-visual clips from a Sensitive Artificial Listener (SAL) task. In this task one participant took the role of the user and another played the role of an embodied conversational agent, using one of the four SAL characters in the SAL system. The laughter in these interactions was largely conversational and social, and was incidental to the task of interacting with the avatar or with a person pretending to be an avatar. In total 443 instances of laughter were extracted from 345 video clips.

9.4.2 Hilarious Laughter Collection

UCL Motion Capture Stick Figure Stimuli

There were a variety of data gathering sessions dedicated to gathering hilarious laughter. One of the goals of these sessions was to gather data related to body movement, so a focus was on motion capture elements. Two data gathering sessions were dedicated to collecting this data. One in Belfast developed the laughter induction techniques [73], and was followed by a similar session conducted at UCL in which the motion capture data was made available as part of the database [39, 72]. The available data consists of 126 animated "stick figure" video stimuli of laughter that has been categorized as either hilarious, social, fake, awkward or not a laugh.

Multimodal Multiperson Corpus of Laughter in Interaction (MMLI)

The MMLI database [86] focused on gathering multimodal full body movement laughter data. This data was collected during recording sessions of French speakers made in Paris. It contains both induced and interactive laughs from human triads. 500 laugh episodes were collected from 16 participants. The data consists of 3D body position information, facial tracking, multiple audio and video channels as well as respiration data.

Belfast StoryTelling Corpus

The Belfast Story-telling corpus was comprised of six sessions of groups of three or four people telling stories to one another in either English or Spanish. The story-telling task was based on the 16 Enjoyable Emotions Induction Task [48]. Participants prepared stories related to each of 16 positive emotions or sensory experiences and were seated around a central table, and each participant wore a head-mounted microphones. HD webcams and depth cameras (Kinect) captured audiovisual streams, facial features, face direction and depth information. Participants took turns at recalling a story with occasional open discussion. Synchronized recording of data streams was achieved using the Social Signal Interpretation (SSI) framework [136].

9.4.3 Conversational Laughter Collection

Although there was much laughter that could be termed conversational laughter within the Belfast Story-telling Database, the activity that the participants were engaged in was not strictly a conversation. To ensure that the database contained laughter that was taken from conversations between people, we devised a very minimal task to capture conversations that were as natural as possible given the presence of cameras, microphones and depth sensors, and between only two participants. There were two versions of this task, one recorded in Belfast, and the other in Peru. Participants were asked to talk on a topic randomly selected from a pre-determined list, but to continue talking freely until they felt a new topic was needed. Sessions lasted for an hour. Some dyads used up to 10 topics in their session, whereas many used only one.

Belfast Conversational Dyads

While the task differed from the one used in the Belfast storytelling database (the random topic task instead of the 16 enjoyable emotions task) the recording set up was almost identical. HD webcams, head-mounted microphones and Kinect sensors were used to record interactions between participants who sat opposite each other. The various data streams were once again synchronized using the SSI software. 10 pairs were recorded.

Peruvian Conversational Dyads

A mobile version of the data capture system was devised and taken to Peru to capture interactions between Peruvian conversational pairs. Unfortunately, the depth sensors could not be made to function, but the same HD webcam and headmounted-microphones were sued to ensure quality recordings. The data streams were synchronized using the SSI software, and 20 interacting dyads were recorded. These recordings involved the speakers interacting in Spanish.

9.4.4   Annotation

An annotation scheme, which had seemed a straightforward endeavor at the outset of the ILHAIRE project, proved to be a much more difficult proposition. We have argued that the inherent ambiguity and underdetermined nature of laughter mean that it does not yield easily to simple a classification system (ref underdetermined). However, within the project we did develop a set of guidelines for the segmentation of laugh episodes; these were used to create a significant amount of annotated material that was then rated along a number of dimensional scales. A substantial amount of laughs have been annotated and rated and this information is available at the ILHAIRE laughter database website.[3] In conclusion, the ILHAIRE laughter database has achieved its goals and now represents the most comprehensive set of laughter resources currently available for use by the research community. It remains an active research database and new materials and annotations continue to be added. We encourage researchers to send us any annotations or data gathered using the database materials so we can incorporate them into the database for the use of other researchers.

9.5   Automatic Detection, Recognition and Characterization

Systems that understand and can also positively impact on social communication require algorithms for the detection and recognition of laughter.

---

[3] http://www.qub.ac.uk/ilhairelaughter.

This can rely on knowledge, models and technologies in sensor systems, signal processing, machine learning, and in particular related to automatic detection and classification of audio-visual signals. Microphone and video cameras can provide raw signals to analyse acoustic and visual modalities. Nowadays, cheap depth cameras are also available through their popularization for use in video gaming. They provide for each pixel a measure of the distance of solid opaque objects from the camera sensor. With these, it is hence possible to get improved accuracy in the capture of 3D cues from the face and the body, where video-only cameras could otherwise fail. In some cases, researchers can also have recourse to more specific equipment, such as respiration sensors, or accurate tracking solutions, for instance based on optical motion capture and gaze tracking.

Then, it is necessary to research on the features to be extracted from the raw sensor signals, on the categories and characteristics that are useful to be recognized and measured, and on the models to be used. This can rely on knowledge about the morphological features of laughter (cf. Sect. 9.3) that really matter.

As machine learning remains a state-of-the-art tool for detection and recognition, also in this area of social signal processing, large annotated corpora containing many instances of the phenomena of interest are also necessary (cf. Sect. 9.4).

Early work essentially focused on designing laughter detectors from audio signals and recognized facial expressions. Such work was actually fueled by research projects starting around the year 2000. At the time, data collection was first achieved through microphones only, sometimes in naturalistic social settings, such as group meetings [53]. Multimodal capture systems including video cameras came a bit later. Since 2010, it is also possible to capture 3D information using cheap hardware, to the benefit of research on leveraging body motion and multimodal approaches. However, this had not been done for laughter analysis before ILHAIRE. Most modalities have actually been further considered, and multimodal approaches for combining them have been proposed. Studies covered detection, characterization of important dimensions such as laughter intensity, and classification of laughter in generic categories. On the side of acoustic analysis, the project came up with novel approaches for making use of automatic recognition of either phonetic or else purely data-driven symbolic units whose local statistics have been used for laughter recognition, yielding beyond state-of-the-art performance. Also, a special focus has been on estimating laughter intensity, one of its most important facet, with first studies on that aspect. Besides, to our knowledge, there was also no previous work on automatic laughter detection and analysis from body movement and respiration signals. The project closed these gaps. Finally, on the side of multimodal integration, a novel approach outperfoming previous proposal of early and late fusion has also been published. The following sections are organized according to the covered modalities.

9.5.1 Acoustic

Early studies on laughter detection often used acoustic features initially designed for speech analysis as well as automatic recognition, classification, statistical modeling and time series modeling, in particular Hidden Markov Models (HMMs) with Gaussian Mixture models (GMMs). Spectral coefficients and HMMs have been used in [65]. In [11], Mel-Frequency Cepstral Coefficients (MFCCs) and perceptual features were applied together with HMMs. In [55], the authors used a different approach relying on Support Vector Machines for classification, still fed with MFCCs though. In [124], the authors used Artificial Neural Networks (ANNs) fed with Perceptual Linear Prediction (PLP) features. In [61], a similar approach but with longer temporal feature windows allowed to reach better performance. Reported results in terms of detection equal error rate (false positive rate and false negative rate) range from about 30 % down to 8 % in [61]. A figure of 3 % has been achieved but using speaker dependent models and pre-segmented laughs.

Within ILHAIRE, such approaches have been applied while novel techniques have also been developed. Although laughter is mostly non-verbal and non-articulated, it nevertheless exhibits some vocal tract configurations close to phonetic sounds. Automatic laughter transcription through phonetic labels has hence been found useful [130]. Following up, in [135], detection has been addressed through a two stages approach. Instead of estimating the probability of laughter using a model based on acoustic features directly as in previous work, it first uses a generic phonetic recognizer to obtain a symbolic sequence. Short audio segments are then characterized by one or two histograms describing the distribution of symbols in the sequence. These constitute feature vectors on which to base a classification model. When combined with more traditional features, it was possible to detect laughter and filler events with 88 % accuracy (unweighted average area under curve (AUC)), hence 4 % absolute above a baseline approach on a published benchmark [116]. A related approach is to use n-gram models of such symbolic sequence to model the patterns of laughter, as evaluated in [92]. A detection F-measure of 75 % has been reached. An additional specificity of that development is to use non-phonetic symbolic units that are defined automatically using the Automatic Language Independent Speech Processing (ALISP) method, hence presenting some potential for improved modeling of non-verbal vocal sounds.

Categories and Intensity

Once laughter is detected, one would want to identify some of its important characteristics. Early results by others on recognizing laughter categories have been obtained in [12], where phonetic features and HMMs were applied to classify among four types (hearty, amused, satirical and social) with an accuracy close to 75 %.

As explained earlier, an important facet to be measured is laughter intensity, which has been shown to correlate with the social versus amused categories, while having the advantage of being continuous instead of categorical. Within ILHAIRE, in [82, 126], a research on the audio and visual cues that can be extracted automatically and correlate with the perception of laughter intensity has been conducted. On the acoustic side, the range of variation of several of the MFCC coefficients, but also of the spectral flux, loudness and pitch were found to be the most important acoustic features, with a correlation coefficient of the best features with laughter intensity ratings reaching 83 %.

9.5.2 Facial Expressions

Recent possibilities for automated tracking of specific facial features or action units have started to be applied to laughter detection and characterization. In [52], spatial locations of facial feature points are tracked using markerless video processing, and used for laughter detection. In [103], principal component analysis of tracked spatial location of feature points is used to obtain features from video signals, and GMMs as well as SVMs were used for classification. In [95, 96], head movement and facial expressions, obtained through facial feature points tracking from the video channel too. These different publications actually applied a multimodal framework, combining acoustic features with visual features reaching detection and classifications accuracies above those obtained when using individual modalities. In [95], an accuracy of 75 % to distinguish three classes, namely unvoiced laughter, voiced laughter and speech, is reported.

Within ILHAIRE, evaluations of face tracking approaches for estimating Facial Animation Parameters (FAPs) and intensities of Action Units (AUs) have been performed [102]. A smile detector based on estimated action units has also been used as one of the component of laughter detection and laughter intensity estimation within a multimodal approach, evaluated in interactive settings.

Categories and Intensity

The previously mentioned works [82, 126] also covered the visual side of intensity estimation. It was shown that the maximum opening of the mouth/jaw as well as the lip height and lower lip protrusion were identified as the most important, with a correlation coefficient of the best features with laughter intensity ratings reaching 68 %. A study of audio and visual features that differ in laughter inhalation and exhalation phases was also proposed. Some features present different patterns finally enabling to distinguish these phases automatically.

9.5.3 Body Movement and Gestures

Body movement during laughter has less background work and available corpora to base the studies on. Although some previous studies described the morphological attributes of laughter, it was still necessary within ILHAIRE to gather more detailed statistics related to specific motion patterns appearing during laughter [40]. These studies used recordings done within the project and including motion capture using high-end optical hardware. Following a preliminary visual inspection, statistics were drawn from laughter segments exhibiting cues from all three modalities (audio, facial and body), hence covering essentially laughter of high intensity (amused emotion). Shoulder shaking has been identified as the most frequent. Then, torso rocking, torso throwing, knee bending and torso leaning were second, each two times less frequent than shoulder shaking. Finally, head shaking and shoulder contraction appeared, but only rarely. Given the importance of torso movement, subsequent research on automatic detection focused essentially on that aspect.

In [66, 132], shoulder movement was tracked either based on a body skeleton extracted from a depth sensor signal, of from shoulder tracking based on machine vision. Features characterizing this motion are then computed, including the correlation between the movement of the two shoulders, the overall kinetic energy, and the periodicity of the movement. Based on that, a system for the automatic detection using commodity hardware (video and depth cameras used in conjunction) has then been proposed and evaluated in [67, 69]. Several motion features were extracted, accounting for shoulder motion correlated with torso (trunk) motion, but also directly torso and head. Evaluation relied on Kohonen self-organising maps, showing significantly above chance estimation results.

Categories and Intensity

The previously referenced work from ILHAIRE [69] also proposed an evaluation of the selected body motion features for automatic estimation of intensity, showing significantly above chance estimation results too.

Another proposal was made earlier [39, 40] within the project where body motion was investigated for laughter type recognition among five categories (hilarious, social, awkward, fake, and non-laughter). Features characterizing hand gesture, shoulder movement, neck/spine bending, as well as kinetic energy of several upper body articulations were extracted. Several classification approaches were compared (k-nearest neighbors, multilayer perceptron, linear and kernel ridge and support vector regression, random forests) with the random forest method yielding the best performance. Tests were made using motion data captured using full-body motion capture equipment. Automatic recognition performance reached about 66 %, which was also shown to approach human rating levels.

In addition to movement directly induced by laughter, the phenomenon may also trigger complementary movement due to the amused state or the need to replace speech with other cues (using pointing, clapping, illustrator, or other gestures). Studies towards understanding these are ongoing.

9.5.4 Respiration and Muscular Activity

Physiological responses such as elevated heart rate can accompany laughter. Also, its production is intimately linked to respiratory patterns, as described in [109]. A distinctive pattern can indeed be observed, consisting in a rapid exhalation followed by a period of smaller exhalations at close-to-minimum lung volume. This pattern, accompanied by contractions of the larynx and epiglottis and facial patterns, result in the specific sound, facial and body movement being observed. On may hence try to detect and characterize laughter directly from respiration measurements, and their underlying muscular activity. In previous work by other [35], myoelectric signals from the diaphragmatic muscle were measured. Detection of laughter was shown to be possible through a threshold-based approach on the amplitude of the high-frequency component of the captured signal. No formal evaluation of the detector itself was proposed however.

Another approach studied in ILHAIRE was to rely on measurements of the thoracic (chest) circumference [132]. Features enabling the detection of laughter from these respiration signals are computed including the following sequence of events: a sharp change in current respiration state, a period of rapid exhalation resulting in rapid decrease in lung volume, a period of very low lung volume. Formal evaluations were performed later, using recognition through HMM models, with a classification accuracy of 69 %, validating the approach (results not published yet).

9.5.5 Multimodal Fusion

Previous studies by others on combining acoustic and visual cues have been published. In [95, 96] (already introduced here above), late fusion (decision level) has been applied using a sum rule or an artificial neural network. In [52], the output of the acoustic and facial detectors are combined with an AND operator, hence a form of late fusion too. Decision level late fusion was also used in [103].

Within ILHAIRE, late fusion of estimations from audio, body and respiration modalities has been implemented in [132] for laughter detection. In [64], an event driven real-time fusion system was proposed. It rather corresponds to a late fusion mechanism, with some additional time-based accumulation. This approach proved particularly robust for the case of laughter detection, since it does not directly fuse identical time frames throughout modalities, but rather computes probabilities indirectly by accumulating shorter, detection-indicating and possibly time-shifted events. Evaluation was performed on an enjoyment detection task (enjoyment defined as an episode of enjoyable emotion, which may hence also include just smile segments). From 54 and 72 % accuracy for audio-based and image-based detection, the fusion approach reached about 79 %, outperforming more traditional early or late fusion schemes (only reaching between 65 and 68 %).

Experimental studies on tuning of laughter analysis to genre and culture were also carried out with promising results. They rely on a range of techniques actually enabling the parameters of detection/classification models to be adapted to specific demographic subgroups or subjects, similarly to what is done in speech synthesis and recognition technology.

Finally, for more information, a survey on multimodal fusion within human-agent dialogue has been proposed by André et al. [1].

9.6 Automatic Generation and Synthesis

Having a proper understanding of the nature of multimodal signals during laughter is necessary, not only to inform on the proper features and models for detection and characterization purposes but also on the proper models for generating signals that sound and look natural, or to enable experimental protocols. Therefore generation and synthesis are covered here.

9.6.1 Acoustic Synthesis

Despite previous work on synthesizing "emotional speech" (cf. review by Burkhardt and Campbell [9]), acoustic laughter synthesis is an almost unexplored domain. In 2007, Sundaram and Narayanan [120] synthesized laughter vowels by Linear Prediction. To obtain the repetitive laughter pattern, they modeled the laughter energy envelope with the equations of an oscillating mass-spring system.

The same year, Lasarcyk and Trouvain [63] compared laughs synthesized by a 3D modeling of the vocal tract and diphone concatenation. The articulatory system gave better results, but synthesized laughs could still not compete with natural human laughter. Beller [5] proposed an original approach to laughter synthesis, as voiced laughter is synthesized from a neutral speech sentence.

Sathya et al. [115] synthesized voiced laughter bouts by controlling several excitation parameters of laughter vowels: pitch period, strength of excitation and amount of frication. After analyzing these features on a range of human laughs, Sathya et al. concluded that the pitch contour and the strength of excitation of laughter calls can be approximated by quadratic functions, while the amount of frication tends to decrease within and across calls.

Cagampan et al. [10] synthesized laughs by concatenating syllables. Laughs were segmented into syllables with different labels denoting laughter vowels ('ha', 'he', 'hi', 'ho', 'hu'), grunt- and snort-like syllables, as well as the laughter onset and offset. These units were then combined to form laughs with four syllables in the apex, plus the possibility of an onset and an offset.

A real-time laughing instrument has been developed by Oh and Wang [90]. Their main objectives were expressivity and control, rather than the quality of the synthesis or laughter naturalness. They synthesized vowels by formant synthesis (source-filter decomposition).

Moreover, recent works from Oh and Wang [89] to modulate speech and make it sound like speech-laugh, as opposed to all previous attempts on pure laughter. The method takes speech as input and segments it into syllables, based on the energy envelope. Then they provide control over several parameters of the syllables that can be affected by laughter: intensity contour, maximum pitch value, tempo regularity (the degree to which segmented speech syllables are fetched to an isochronous tempo), rhythm (the periodicity of syllables between 4 and 6 Hz).

Within ILHAIRE an HMM-based approach was utilized. This method is based on a framework which became popular in the field of speech synthesis in the last decade. In HMM-based parametric speech synthesis, the spectrum, F0 and duration of phonemes are modeled in a unified framework [137]. Based on the resulting HMM, a maximum-likelihood parameter generation algorithm is used to predict the source/filter features [123], which are then sent to a parametric synthesizer to produce the waveform. Urbain et al. exploited this technique to perform HMM-based acoustic laughter synthesis [125, 128, 130, 131, 134]. They investigated the synthesis audio laughter from arousal curves by comparing the arousal curve given as an input to the arousal curves of laughter syllables that were available in the database. The transcriptions of the best matching syllable were used to drive the HMM-based acoustic laughter synthesis system. In [6], Bollepalli et at. they compared the use of different vocoders for the specific purpose of acoustic laughter synthesis. Using the same approach, "speech-laugh", namely the phenomenon of laughter occurring at the same time as speech by intermingling with it or by interrupting it, an HMM-based speech-laugh synthesis system has been developed by El Haddad et al. [32]. This system involves first creating HMM models of laughter and speech-smile. Then, some vowels in the synthesized speech-smile sentences are replaced by laughter bursts.

9.6.2 Visual Synthesis

As for the synthesis of laughter sound, very few studies exist regarding the synthesis of laughter facial and body behaviour. Since the few existing studies focused either only on facial synthesis or else only on body synthesis, this section is organized as a chronological survey of studies which are related to visual laughter synthesis without further categorization.

In 2008, a parametric physical chest model, which could be animated from laughter audio signals, was proposed by DiLorenzo et al. [21]. The model is able to produce realistic upper body animation but facial animation is not addressed.

The next year, Cosker et al. [16] studied non-verbal articulations including laughter. They explored the possible mapping between facial expressions and their related audio signals. Hidden Markov Models (HMMs) were used to model the audio-visual correlation. As with DiLorenzo et al., the animation was audio-driven.

Further research has been pursued within ILHAIRE. In 2012, Niewiadomski et al. compared three possible approaches regarding visual laughter animation [84]. The same year, Niewiadomski and Pelachaud [80] considered how laughter intensity modulates facial motion. A specific threshold is defined for each key point. Each key point moves linearly according to the intensity if it is higher than the corresponding threshold. So, if the intensity is high, the facial key points concerning laughter move more. In this model, facial motion position depends only on laughter intensity.

More recently, in 2013, two studies [85, 132] included the use of laughter capable avatars for human-machine interactions. Two different avatars animated from recorded data were proposed. One of them is the Greta Realizer [83] which takes as controls either high level commands using the Facial Action Coding System (FACS) or low level commands using Facial Animation Parameters (FAPs) of the mpeg-4 standard for facial animation. Greta generates an animation corresponding to the initially recorded laughter (copy-synthesis). The other avatar is the Living Actor[4] which plays a set of manually drawn animations.

A recent study published in 2014 aiming at synthesizing facial laughter was proposed by Çakmak et al. [14]. The approach followed was to model facial deformations by means of landmark trajectories. The basic steps followed throughout the work are: recording of the 3D data using a motion capture system, post-processing to shape the data for training, training of Hidden Markov Models (HMMs) on this data, synthesizing trajectories based on this data and retargeting the synthesized trajectories on a 3D face model to finally render a video output. To be able to build accurate models, this study needed the building of a specific audio-visual laughter database containing synchronous audio and 3D motion capture data in sufficient amount for a single subject [13].

The same year, Ding et al. [22, 24] developed a generator for face and body motions that takes as the input the sequence of pseudo-phonemes and their duration [22, 23]. Lip and jaw movements are further driven by laughter prosodic features that are based on a contextual Gaussian Models approach. The relationship between input data (pseudo-phonemes and acoustic features) and motion features is first modeled and then the model is used to produce laughter in real-time. Head an eyebrow generation is based on the selection and concatenation of motion segments from the database. Torso and shoulder motions are derived from head motion.

Another study in 2014 by Niewiadomski et al. [87] propose a procedural method to synthesize rhythmic body movements of laughter based on spectral analysis of laughter episodes. For this purpose, they analyzed laughter body motions from motion capture data and reconstructed them with appropriate harmonics.

Niewiadomski and Pelachaud [81] studied the identification and perception of facial action units displayed alone as well as the meaning decoding and perception of full-face synthesized expressions of laughter. They focused on three factors that may influence the identification and perception of single actions and full-face expressions: their presentation mode (static versus dynamic), their intensity, and the presence of wrinkles. They used a hybrid approach for animation synthesis that combines data-driven and procedural animations with synthesized wrinkles generated using a bump mapping method.

9.7 Interaction Modeling

Enabling conversational agents with laughter capabilities is not only about being able to recognize and synthesize audio-visual laughter signals. It is also concerned by an appropriate management of laughter during the interaction. There is thus a

[4]http://www.cantoche.com/.

need for a laughter-enabled interaction manager (IM), able to decide when and how to laugh so that it is appropriate in the conversation. Despite the body of work in so-called dialog modeling, there was no previous work specifically adressing laughter as indeed, previous research was essentially focused on verbal social communication.

It remains uneasy to define what an appropriate moment to laugh is. This can be seen as a decision making process. These decisions have to be taken according to the interaction context which can be inferred from laughter, speech and smile detection modules (detecting social signals) but also by the task context (for example, if the human is playing a game with the agent, what is the status of the game). Formally, the IM is thus a module implementing a mapping between interactional contexts and decisions. Lets call this mapping a policy.

Describing the optimal policy of the agent is a very tricky task. It would require the perfect knowledge of rules prevailing to the generation of laughter by humans. Interpreting sources of laughter or predicting laughter from a cognitive or psychology perspective is non-trivial. Therefore, a data-driven method has been preferred in the ILHAIRE project. Especially, we adopted a Learning from Demonstrations (LfD) framework to learn the IM policy. Indeed, humans are implementing such a policy and they can provide examples of natural behaviors.

LfD is a paradigm in which an artificial agent learns by observing another agent (artificial or human) performing optimally the task at sight. Several generic methods can be used to implement this paradigm among which two have been explored within the ILHAIRE project: (1) Imitation Learning (IL); (2) Inverse Reinforcement Learning (IRL).

Imitation learning reduces the problem of learning the optimal policy to a classification problem. Indeed, one can see the policy as the result of a process that assigns a decision to an interactional contexts which is similar to the standard classification problem consisting in assigning labels to inputs. For instance, a simple K-nearest neighbors algorithm has been used in [85]. Taking as input the results of audio-visual processing systems able to detect speech and laughter of other participants, this method generated a laughter/silence decision every 200 ms. Yet, classification algorithms usually underperform when trained on unbalanced data. This is the case with laughter which is way less frequent than speech and silence in human-human interactions. For this reason, we introduced a structured-classification method in [98], enabling to emphasize more on laughter and improve classification performances.

Inverse Reinforcement Learning considers laughter as a sequential decision making process. In this framework, the decisions taken by the conversational agent at a given time are supposed to have an impact on the reactions of the user(s) right after or even later and, so, on the course of the interaction. Therefore, the optimization of this module should take the whole interaction into account, including the impact of local decisions onto the future. Sequential decision making processes are generally addressed under the Reinforcement Learning (RL) paradigm in the machine learning literature. In RL, the agent is assumed to optimize a cumulative function of immediate rewards, provided after every decision. Because it learns to maximize the sum of rewards, it learns a sequence of decisions instead of local decisions as in imitation learning. Yet, the problem of defining the reward remains. IRL is a method by which, observing an expert agent, another agents learns the unknown reward the expert is optimizing. This is unfortunately an ill-posed problems. Indeed, the null reward makes any decision policy optimal. In addition, IRL often requires observing expert and non-expert data which is obviously impossible in laughter studies (what is a non-expert laughter?). During the ILHAIRE project, we thus developed innovative IRL algorithms that could learn non-trivial rewards from expert-only data [59, 60] and applied them successfully to laughter/silence decision in the final project demonstrator.

9.8   Application Perspectives

Many applications could benefit from results of research on human laughter. The applications mentioned here stem from the ILHAIRE project results and from previous research. First, the automatic detection of laughter episodes and the extraction of their characteristics could be useful for indexing and searching speech and conversational databases, such as meeting recordings. Laughter is often related to amusement, so its detection could help in detecting jokes. Laughter also has a role as regulator of conversations and is a marker of dialog turns and topic changes. Its detection could hence facilitate automatic speaker diarization and topic segmentation. Laughter also affects the perception of emotional valence, arousal and dominance. Effective laughter detection could hence provide an additional layer of automatic enriched transcriptions, one that would help infer speaker roles and moods. The interpretation of responses to laughter would also offer information about the emotional state and personality traits of the subjects. There are indeed demonstrated inter-individual differences in the way laughter is perceived and what kind of feedback expressions it triggers in return.

The integration of "laughing skills" into interactive systems (such as robots, virtual conversational agents, computer mediated inter-personal communication systems) would offer several other benefits. Laughter would not only make interaction with such systems more natural, it would also offer an additional mean for computers to influence the course of conversations. Using computers to regulate multi-party conversations/meetings has already been suggested in [105]. This remains to be researched whether such systems will face negative perceptions (similar to an uncanny valley effect) and how much work will be necessary to bridge that gap if any. Besides, ethical issues will need to be considered.

Being a frequent natural indicator of positive moods and emotions (amusement but also other enjoyable emotions), such laughing machines would be better equipped to align their expressive behaviour to the emotional color of the interaction. This would lead to increased naturalness and impact. Positive mood changes induced by conversational systems using laughter have been demonstrated within ILHAIRE as well as by others, in media consumption setups [35, 46, 85, 129], in "affective" installations [78, 93, 118], and in games played with virtual conversational agents [68]. More specifically, when a virtual agent was interactively responding to laughter, a level of contagion associated with spontaneous laughter as well as exhilaration was experienced by the users (Niewiadomski et al. [85], Hofmann et al. [46]). Actually, it has been shown that some properties of laughter trigger different perceptions of valence and arousal. A system able to generate laughter from a more varied repertoire would hence be better equipped to trigger different reactions in return. Here again, ethical issues will need to be considered, especially those related to the flip side of laughter when it is associated with negative feelings (such as ridicule).

Others [34] have also studied some of the positive effects of laughter on learning. Laughter could indeed contribute to improved motivation and self-image in various applications, such as learning, rehabilitation or fitness exercises. In ILHAIRE, a virtual tutor equipped with laughter increased the positive perception of the learning experience (results not published yet).

Laughter is also an important marker of social bonding. It reduces the sense of threat and facilitates cooperation in a group [27, 91]. It could hence be used within computer mediated inter-personal communication systems to favour collaborative and positive outcomes, and as a complement to the inter-personal skills of humans. Laughter, together with its accompanying body postures and movements, can also convey sexual bonding messages that range from solicitation to aversion, depending on which and how many different signals are present [37]. Novelists and screenwriters may be better armed to speculate about the use of such functions by computer systems.

Empathy is a significant and particularly interesting component of social bonding. Recent experimental results support the hypothesis that anthropomorphism positively affects the empathy of people towards robots. This was tested in [104], where Riek et al. used robots (along a chosen anthropomorphic range) shown to be experiencing mistreatment by humans, hence targeting the pro-social behavioral component of empathy (i.e. the component of empathy that leads to helping behavior toward others in need). Although those experiments were only covering the physical appearance of anthropomorphism, the authors speculate that these results are in fact compatible with Simulation Theory, which states that people mentally simulate the situation of other agents in order to understand their mental and emotive states. Being so pervasive, laughter is also likely to encourage humans to empathize more easily with virtual agents and robots.

In all these applications, the underlying social context has a strong influence. Given the interpersonal differences in perception and feedback to laughter, person-ality traits constitute a crucial contextual element. Analyzing reactions to laughter could hence offer information that would enable systems to better understand and adapt to individual personality. Gelotophobes constitute a specific group that has been studied within ILHAIRE. They exhibit a fear of being laughed at and display negative expressions in reaction. This constitutes a significant social handicap given the pervasiveness of laughter in society. Another opportunity to be pursued in the future would hence be to build interactive systems and interventions designed to help these people. They could participate more frequently in social activities through vir-tual agents making use of laughter expressions that are not perceived negatively. They could also combat their fear through gradual habituation to natural expressions of laughter.

We believe these applications will be facilitated by technologies for the recognition, understanding, interaction modeling, generation and synthesis of a wide range of laughter expressions. These need to be adaptable to varied social contexts, contents, moods and personalities. Also, these should be based on strong knowledge in psychology and machine learning.

9.9   Research Perspectives

Bringing together the fields of engineering and psychology can only strengthen each other. The psychological impact of human-computer interaction systems should be considered as a key metric for the evaluation of their success. However, safeguards must be taken in order to benefit from extrapolating the information gathered. As discussed in [99], this is especially true when evaluating humorous laughter, as many aspects of humour revolve around individual differences. Indeed, there are personal preferences in the type of humour an individual finds funny. Also, as discussed earlier, personality traits matter a lot. Researchers should hence continue to take care of formalizing the experimental protocols and environments as much as possible. Further research on the sense of humour is of particular interest too [108].

Future research should also concentrate on eliciting laughter in natural situations and study its morphology, subjective experience, and social consequences. The studies should also aim towards the inclusion of multiple modalities in laughter coding. Here, the respiration and body movements as well as the interplay of their features may be considered. Future works should also attempt to arrive at a classification of laughter that spans over more than one modality.

Regarding automatic synthesis, the quality achieved with HMM-based systems is significantly higher than with other laughter synthesis approaches. Such synthetic laughs nevertheless remain far from actual human laughs. Future work needs to focus on improving the naturalness, and on the synchronization between the audio and visual modalities, as shown in a first exploratory study by Çakmak et al. [15]. Other aspects include the level of control and diversity of laughter categories and characteristics, and the desired reactivity (hence the ability to change the desired synthesis output on the fly, adapting to a change in context or interaction state). Studies [17, 33] focusing on reactive laughter synthesis provided interesting results. Recent research on increasing the diversity of laughter that can be produced by the computer [30, 31] also yielded promising results.

Laughter-enabled interaction management remains a tricky task even though Learning from Demonstrations (LfD) proved to be a promising way to handle it. Yet it requires a lot of annotation work, new data collection campaigns for each task, integration of non-trivial contextual information. To address these issues, machine learning offers several perspectives that can be envisioned among which semi-supervised learning to train the models from little amounts of labeled data, transfer learning to capitalize knowledge acquired from task to task and automatic feature selection combined with non-parametric methods to scale with the dimensions of the input space.

The Imitation Learning (IL) and Inverse Reinforcement Learning (IRL) algorithms developed within the ILHAIRE project could be easily extended to implement most of these methods. On another hand, even though IRL considers the laughter/silence decision process as a sequential decision process, it only models the long-term influence of the artificial agent on the interaction. Another direction of research would therefore be to model the interaction as a multi-agent problem where the human users would also try to influence the agents behavior (like trying to generate contagion to the agent).

9.10 Conclusions

Laughter will be a key component of "future socially believable behaving systems". This chapter reviewed recent work on endowing computer systems with the ability to master this crucial social signal, in humorous as well as more general conversational contexts. It covered a range of fundamental and technological topics including: (1) studies on the roles, functions and perception of laughter, and (2) technological developments on recognizing and synthesizing laughter, and on modeling interactions including laughter.

Research on laughter in human communication and on its use by computers is currently gaining in popularity. A comprehensive review within the limits of a book chapter is already becoming difficult to achieve. Here, we focused essentially on the work done within the ILHAIRE project, together with what we identified as pioneering work by others. We invite the readers to continue their exploration through recent publications by us and by other researchers active in this area. One may look in particular for publications by Pelachaud, André, Camurri, Berthouze (participants in ILHAIRE), as well as Devillers, Campbell, Trouvain, Truong, and numerous other researchers we can not cover here. We hope this chapter will be the starting point for a fascinating journey, back in time through the evolutionary roots of laughter, amusement and social behaviour; and into the future of people engaging with digital media.

Acknowledgments We would like to acknowledge all colleagues within the ILHAIRE project, from the following partner organisations: University of Mons (Belgium), Télécom ParisTech / Centre National de la Recherche Scientifique (France), University of Augsburg (Germany), Università degli Studi of Genova (Italy), University College London (United Kingdom), Queen's University 'Belfast (United Kingdom), University of Zurich (Switzerland), Supélec (France), Cantoche (France), University of Lille (France). Our thanks go to Laurent Ach, Elisabeth André, Hane Aung, Emeline Bantegnie, Tobias Baur, Nadia Berthouze, Antonio Camurri, Gerard Chollet, Roddy Cowie, Will Curran, Yu Ding, Stéphane Dupont, Thierry Dutoit, Matthieu Geist, Harry Griffin, Jing Huang, Jennifer Hofmann, Florian Lingenfelser, Anh Tu Mai, Maurizio Mancini, Gary McKeown, Benoît Morel, Radoslaw Niewiadomski, Sathish Pammi, Catherine Pelachaud, Olivier Pietquin, Bilal Piot, Tracey Platt, Bingqing Qu, Johannes Wagner, Willibald Ruch, Abhisheck Sharma, Lesley Storey, Jérôme Urbain, Giovanna Varni, Gualtiero Volpe, and their colleagues and co-authors. They all contributed to the initial ideas, to the teambuilding, or to the scientific/research developments within the project. The research leading to these results has received funding from the EU Seventh Framework Programme (FP7/2007–2013) under grant nbr. 270780 (ILHAIRE project).

References

1. André E, Martin JC, Lingenfelser F, Wagner J (2013) Multimodal fusion in human-agent dialogue. In: Rojc M, Campbell N (eds) Coverbal synchrony in human-machine interaction. CRC Press, Boca Raton
2. Bachorowski JA, Owren MJ (2001) Not all laughs are alike: voiced but not unvoiced laughter readily elicits positive affect. Psychol Sci 12(3):252–257
3. Bachorowski JA, Owren MJ (2003) Sounds of emotion. Ann N Y Acad Sci 1000:244–265
4. Bachorowski, J.A., Smoski, M.J., Owen, M.J.: The acoustic features of human laughter. J Acoust Soc Am 110(3, Pt1), 1581–1597 (2001)
5. Beller G (2009) Analysis and generative model for expressivity. Applied to speech and musical performance. PhD thesis, Université Paris VI Pierre et Marie Curie
6. Bollepalli B, Urbain J, Raitio T, Gustafson J, Cakmak H (2014) A comparative evaluation of vocoding techniques for hmm-based laughter synthesis. In: 2014 IEEE international conference on acoustics, speech and signal processing (ICASSP), pp 255–259. doi:10.1109/ICASSP. 2014.6853597
7. Bonin F, Campbell N, Vogel C (2012) Laughter and topic changes: temporal distribution and information flow. In: CogInfoCom 2012–3rd IEEE international conference on cognitive info communications. Kosice, Slovakia, pp 53–58
8. Bryant GA, Aktipis CA (2014) The animal nature of spontaneous human laughter. Evol Hum Behav 35(4):327–335
9. Burkhardt F, Campbell N (2015) Emotional speech synthesis. In: Calvo R, D'Mello S, Gratch J, Kappas A (eds) The oxford handbook of affective computing. Oxford University Press, Oxford
10. Cagampan B, Ng H, Panuelos K, Uy K, Cu J, Suarez M (2013) An exploratory study on naturalistic laughter synthesis. In: Proceedings of the 4th international workshop on empathic computing (IWEC'13). Beijing, China
11. Cai R, Lu L, Zhang HJ, Cai LH (2003) Highlight sound effects detection in audio stream. In: Proceedings of the 2003 international conference on multimedia and expo, 2003. ICME '03, vol 3, pp III-37–40. doi:10.1109/ICME.2003.1221242
12. Campbell N, Kashioka H, Ohara R (2005) No laughing matter. In: Proceeding of INTERE-SPEECH, pp 465–468. Lisbon, Portugal (2005)
13. Çakmak H, Urbain J, Dutoit T (2014) The AV-LASYN database: a synchronous corpus of audio and 3D facial marker data for audio-visual laughter synthesis. In: Proceedings of the 9th international conference on language resources and evaluation (LREC'14)
14. Çakmak H, Urbain J, Tilmanne J, Dutoit T (2014) Evaluation of HMM-based visual laughter synthesis. 2014 IEEE international conference on acoustics speech and signal processing (ICASSP). IEEE, Florence, pp 4578–4582
15. Çakmak H, Urbain J, Dutoit T (2015) Synchronization rules for HMM-based audio-visual laughter synthesis. In: 2015 IEEE international conference on acoustics speech and signal processing (ICASSP). IEEE, South Brisbane, pp 2304–2308
16. Cosker, D., Edge, J.: Laughing, crying, sneezing and yawning: automatic voice driven animation of non-speech articulations. In: Computer animation and social agents (CASA) (2009)
17. dAlessandro N, Tilmanne J, Astrinaki M, Hueber T, Dall R, Ravet T, Moinet A, Cakmak H, Babacan O, Barbulescu A, Parfait V, Huguenin V, Kalayc ES, Hu Q (2014) Reactive statistical mapping: towards the sketching of performative control with data. In: Rybarczyk Y, Cardoso T, Rosas J, Camarinha-Matos L (eds) Innovative and creative developments in multimodal interaction systems, IFIP advances in information and communication technology, vol 425, pp 20–49. Springer, Heidelberg (2014)
18. Davila Ross M, Owren MJ, Zimmermann E (2009) Reconstructing the evolution of laughter in great apes and humans. Current Biol 19(13):1106–1111
19. Davila Ross M, Allcock B, Thomas C, Bard KA (2011) Aping expressions? chimpanzees produce distinct laugh types when responding to laughter of others. Emotion 11(5):1013–1020

20. Devillers L, Vidrascu L (2007) Positive and negative emotional states behind the laughs in spontaneous spoken dialogs. In: Interdisciplinary workshop on the phonetics of laughter, p 37
21. DiLorenzo P, Zordan V, Sanders B (2008) Laughing out loud: control for modeling anatomically inspired laughter using audio. ACM Trans Graph
22. Ding Y (2014) Data-driven expressive animation model of speech and laughter for an embodied conversational agent. PhD thesis, Télécom ParisTech (2014)
23. Ding Y, Huang J, Fourati N, Artières T, Pelachaud C (2014) Upper body animation synthesis for a laughing character. In: Intelligent virtual agents. Springer, Heidelberg, pp 164–173
24. Ding Y, Prepin K, Huang J, Pelachaud C, Artières T (2014) Laughter animation synthesis. In: Proceedings of the 2014 international conference on Autonomous agents and multi-agent systems. International foundation for autonomous agents and multiagent systems, pp. 773–780
25. Douglas-Cowie E, Campbell N, Cowie R, Roach P (2003) Emotional speech: towards a new generation of databases. Speech Commun 40(1–2):33–60. doi:10.1016/S0167-6393(02)00070-5. http://www.sciencedirect.com/science/article/pii/S0167639302000705
26. Douglas-Cowie E, Cowie R, Sneddon I, Cox C, Lowry O, McRorie M, Martin JC, Devillers L, Abrilian S, Batliner A, Amir N, Karpouzis K (2007) The humaine database: addressing the collection and annotation of naturalistic and induced emotional data. In: Paiva A, Prada R, Picard R (eds) Affective computing and intelligent interaction, Lecture notes in computer science, vol 4738. Springer, Heidelberg, pp 488–500
27. Dunbar R (2008) Mind the gap: or why humans are not just great apes. In: Proceedings of the British academy, vol 154. Joint British academy/British psychological society annual lecture
28. Ekman P (2003) Sixteen enjoyable emotions. Emotion Res 18(2):6–7
29. Ekman P, Friesen WV, Hager JC (2002) Facial action coding system: a technique for the measurement of facial movement
30. El Haddad K, Çakmak H, Dupont S, Dutoit T (2015) Towards a speech synthesis system with controllable amusement levels. In: Proceedings of 4th interdisciplinary workshop on laughter and other non-verbal vocalisations in speech. Enschede, The Netherlands
31. El Haddad K, Dupont S, d'Alessandro N, Dutoit T (2015) An HMM-based speech-smile synthesis system: an approach for amusement synthesis. In: Proceedings of 3rd international workshop on emotion representation, analysis and synthesis in continuous time and space (EmoSPACE15). Ljubljana, Slovenia
32. El Haddad K, Dupont S, Urbain J, Dutoit T (2015) Speech-laughs: an HMM-based approach for amused speech synthesis. In: International conference on acoustics, speech and signal processing (ICASSP 2015)
33. El Haddad K, Moinet A, Çakmak H, Dupont S, Dutoit T (2015) Using mage for real time speech-laugh synthesis. In: Proceedings of 4th interdisciplinary workshop on laughter and other non-verbal vocalisations in speech. Enschede, The Netherlands
34. Fredrickson B (2004) The broaden-and-build theory of positive emotions. Philos Trans R Soc B Biol Sci 359:1367–1378
35. Fukushima S, Hashimoto Y, Nozawa T, Kajimoto H (2010) Laugh enhancer using laugh track synchronized with the user's laugh motion. In: CHI '10 extended abstracts on human factors in computing systems, CHI EA '10, pp 3613–3618. ACM, New York. doi:10.1145/1753846.1754027
36. Glenn PJ (2003) Laughter in interaction. The discourse function of laughter in writing tutorials. Cambridge University Press, Cambridge
37. Grammer K (1990) Strangers meet: Laughter and nonverbal signs of interest in opposite-sex encounters. J Nonverbal Behav 14(4):209–236. doi:10.1007/BF00989317
38. Greengross G, Miller GF (2011) Humor ability reveals intelligence, predicts mating success, and is higher in males. Intelligence 39(4):188–192
39. Griffin H, Aung M, Romera-Paredes B, McLoughlin C, McKeown G, Curran W, Bianchi-Berthouze N (2013) Laughter type recognition from whole body motion. In: 2013 Humaine association conference on affective computing and intelligent interaction (ACII), pp 349–355. doi:10.1109/ACII.2013.64

40. Griffin H, Aung M, Romera-Paredes B, McLoughlin C, McKeown G, Curran W, Berthouze N (2015) Perception and automatic recognition of laughter from whole-body motion: continuous and categorical perspectives. IEEE transactions on affective computing, PP(99). doi:10.1109/TAFFC.2015.2390627

41. Hatfield E, Cacioppo JT, Rapson RL (1994) Emotional contagion. Cambridge University Press, New York

42. Hofmann J (2014) Intense or malicious? the decoding of eyebrow-lowering frowning in laughter animations depends on the presentation mode. Front Psychol 5:1306

43. Hofmann J (2014) Smiling and laughter in positive emotions: personality influences and expressive features. PhD thesis, University of Zurich

44. Hofmann J, Platt T, Ruch W, Proyer RT (2015) Individual differences in gelotophobia predict responses to joy and contempt. Sage Open 5(2):1–12

45. Hofmann J, Platt T, Ruch W, More than amusement: Laughter and smiling in positive emotions (under review)

46. Hofmann J, Platt T, Ruch W, Niewiadomski R, Urbain J (2015) The influence of a virtual companion on amusement when watching funny films. Motiv Emot 39(3): 434–447

47. Hofmann J, Ruch W (2016) Schadenfreude laughter. Semiotika (Special Issue on Laughter)

48. Hofmann J, Stoffel F, Weber A, Platt T (2011) The 16 enjoyable emotions induction task (16-EEIT)—unpublished research instrument, Technical report, University of Zurich, Switzerland

49. Hofmann J, Ruch W, Platt T (2012) The en-and decoding of schadenfreude laughter. sheer joy expressed by a duchenne laugh or emotional blend with a distinct morphological expression? In: Interdisciplinary workshop on laughter and other non-verbal vocalisations in speech proceedings, pp 26–27

50. Holt E (2010) The last laugh: shared laughter and topic termination. J Pragmat 42(6):1513–1525

51. Hudenko WJ, Magenheimer MA (2011) Listeners prefer the laughs of children with autism to those of typically developing children. Autism 16(6):641–655. doi:10.1177/1362361311402856

52. Ito A, Wang X, Suzuki M, Makino S (2005) Smile and laughter recognition using speech processing and face recognition from conversation video. In: Proceedings of the 2005 international conference on cyberworlds, CW '05, pp 437–444. IEEE Computer Society, Washington. doi:10.1109/CW.2005.82

53. Janin A, Baron D, Edwards J, Ellis D, Gelbart D, Morgan N, Peskin B, Pfau T, Shriberg E, Stolcke A, Wooters C (2003) The ICSI meeting corpus. In: 2003 IEEE international conference on acoustics, speech, and signal processing, 2003. proceedings. (ICASSP '03), vol 1, pp I-364-I-367. doi:10.1109/ICASSP.2003.1198793

54. Kayyal M, Widen S, Russell J (2015) Context is more powerful than we think: contextual cues override facial cues even for valence. Emotion 15(3):287–291

55. Kennedy L, Ellis D (2004) Laughter detection in meetings. In: NIST ICASSP 2004 meeting recognition workshop. Montreal, Canada, pp 118–121

56. Kipper S, Todt D (2001) Variation of sound parameters affects the evaluation of human laughter. Behaviour 138(9):1161–1178

57. Kipper S, Todt D (2003) Dynamic-acoustic variation causes differences in evaluations of laughter. Percept Motor Skills 96(3):799–809

58. Kipper S, Todt D (2003) The role of rhythm and pitch in the evaluation of human laughter. J Nonverbal Behav 27(4):255–272

59. Klein E, Geist M, Piot B, Pietquin O (2012) Inverse reinforcement learning through structured classification. In: Bartlett P, Pereira FCN, Burges CJC, Bottou L. Weinberger KQ (eds.) Advances in neural information processing systems 25, pp 1016–1024. URL http://books.nips.cc/papers/files/nips25/NIPS2012_0491.pdf

60. Klein E, Piot B, Geist M, Pietquin O (2013) A cascaded supervised learning approach to inverse reinforcement learning. In: Blockeel H, Kersting K, Nijssen S, Zelezny F (eds) Proceedings of the European conference on machine learning and principles and practice of

knowledge discovery in databases (ECML/PKDD 2013), Lecture notes in computer science, vol 8188, pp 1–16. Springer, Prague (Czech Republic) (2013). URL http://www. ecmlpkdd2013.org/wp-content/uploads/2013/07/327.pdf

61. Knox MT, Mirghafori N (2007) Automatic laughter detection using neural networks. In: INTERSPEECH 2007, 8th annual conference of the international speech communication association, ISCA. Antwerp, Belgium, August 27–31, 2007, pp 2973–2976

62. Kori S (1989) Perceptual dimensions of laughter and their acoustic correlates. Proc Int Conf Phon Sci Tallinn 4:255–258

63. Lasarcyk E, Trouvain J (2007) Imitating conversational laughter with an articulatory speech synthesis. In: Proceedings of the interdisciplinary workshop on the phonetics of laughter. Saarbrücken, Germany, pp 43–48

64. Lingenfelser F, Wagner J, André E, McKeown G, Curran W (2014) An event driven fusion approach for enjoyment recognition in real-time. In: Proceedings of the ACM international conference on multimedia, MM '14. ACM, New York, pp 377–386. doi:10.1145/2647868. 2654924

65. Lockerd A, Mueller FM (2002) Lafcam: leveraging affective feedback camcorder. In: CHI '02 Extended abstracts on human factors in computing systems, CHI EA '02. ACM, New York, pp 574–575. doi:10.1145/506443.506490

66. Mancini M, Varni G, Glowinski D, Volpe G (2012) Computing and evaluating the body laughter index. In: Salah A, Ruiz-del Solar J, Merili E, Oudeyer PY (eds) Human behavior understanding, Lecture notes in computer science, vol 7559. Springer, Heidelberg, pp 90–98

67. Mancini M, Hofmann J, Platt T, Volpe G, Varni G, Glowinski D, Ruch W, Camurri A (2013) Towards automated full body detection of laughter driven by human expert annotation. In: 2013 Humaine association conference on affective computing and intelligent interaction (ACII). IEEE, New Jersey, pp 757–762

68. Mancini M, Ach L, Bantegnie E, Baur T, Berthouze N, Datta D, Ding Y, Dupont S, Griffin H, Lingenfelser F, Niewiadomski R, Pelachaud C, Pietquin O, Piot B, Urbain J, Volpe G, Wagner J (2014) Laugh when you're winning. In: Rybarczyk Y, Cardoso T, Rosas J, Camarinha-Matos L (eds) Innovative and creative developments in multimodal interaction systems, IFIP Advances in information and communication technology, vol 425. Springer, Heidelberg, pp 50–79

69. Mancini M, Varni G, Niewiadomski R, Volpe G, Camurri A (2014) How is your laugh today? In: Proceedings of the extended abstracts of the 32nd annual ACM conference on human factors in computing systems, CHI EA '14. ACM, New York, pp. 1855–1860. doi:10.1145/ 2559206.2581205

70. Matsusaka T (2004) When does play panting occur during social play in wild chimpanzees? Primates J Primatol 45(4):221–229

71. McKeown G, Cowie R, Curran W, Ruch W, Douglas-Cowie E (2012) Ilhaire laughter database. In: Proceedings of the LREC workshop on corpora for research on emotion sentiment and social signals (ES 2012). European language resources association (ELRA), Istanbul

72. McKeown G, Curran W, Kane D, Mccahon R, Griffin HJ, McLoughlin C, Bianchi-Berthouze N (2013) Human perception of laughter from context-free whole body motion dynamic stimuli. In: 2013 Humaine association conference on affective computing and intelligent interaction, pp 306–311. doi:http://doi.ieeecomputersociety.org/10.1109/ACII.2013.57

73. McKeown G, Curran W, McLoughlin C, Griffin H, Bianchi-Berthouze N (2013) Laughter induction techniques suitable for generating motion capture data of laughter associated body movements. In: Proceedings of the 2nd international workshop on emotion representation, analysis and synthesis in continuous time and space (EmoSPACE) In conjunction with the IEEE FG. Shanghai, China

74. McKeown G, Sneddon I, Curran W (2015) Gender differences in the perceptions of genuine and simulated laughter and amused facial expressions. Emot Rev 7(1):30–38

75. McKeown G, Sneddon I, Curran W (2015) The underdetermined nature of laughter. In preparation

76. McKeown GJ (2013) The analogical peacock hypothesis: the sexual selection of mind-reading and relational cognition in human communication. Rev Gen Psychol 17(3):267–287

77. McKeown G, Valstar M, Cowie R, Pantic M, Schroder M (2012) The semaine database: annotated multimodal records of emotionally colored conversations between a person and a limited agent. IEEE Trans Affect Comput 3(1):5–17. doi:10.1109/T-AFFC.2011.20

78. Melder WA, Truong KP, Uyl MD, Van Leeuwen DA, Neerincx MA, Loos LR, Plum BS (2007) Affective multimodal mirror: Sensing and eliciting laughter. In: Proceedings of the international workshop on human-centered multimedia, HCM '07. ACM, New York, pp. 31–40. doi:10.1145/1290128.1290134

79. Miller GF (2001) The mating mind. Vintage, London

80. Niewiadomski R, Pelachaud C (2012) Towards multimodal expression of laughter. In: Intelligent virtual agents. Springer, New York, pp 231–244

81. Niewiadomski R, Pelachaud C (2015) The effect of wrinkles, presentation mode, and intensity on the perception of facial actions and full-face expressions of laughter. ACM Trans Appl Percept (TAP) 12(1):2

82. Niewiadomski R, Urbain J, Pelachaud C, Dutoit T (2012) Finding out the audio and visual features that influence the perception of laughter intensity and differ in inhalation and exhalation phases. In: proceedings of the 4th International workshop on Corpora for research on emotion, sentiment and social signals, satellite of LREC 2012, Istanbul, Turkey

83. Niewiadomski R, Obaid M, Bevacqua E, Looser J, Anh LQ, Pelachaud C (2011) Cross-media agent platform. In: Proceedings of the 16th international conference on 3D web technology. ACM, New York, pp 11–19

84. Niewiadomski R, Pammi S, Sharma A, Hofmann J, Platt T, Cruz R, Qu B (2012) Visual laughter synthesis: initial approaches. In: Interdisciplinary workshop on laughter and other non-verbal vocalisations in speech, Dublin, Ireland

85. Niewiadomski R, Hofmann J, Urbain J, Platt T, Wagner J, Piot B, Çakmak H, Pammi S, Baur T, Dupont S, Geist M, Lingenfelser F, McKeown G, Pietquin O, Ruch W (2013) Laugh-aware virtual agent and its impact on user amusement. In: Proceedings of the international conference on autonomous agents and multi-agent systems, AAMAS (2013)

86. Niewiadomski R, Mancini M, Baur T, Varni G, Griffin H, Aung MSH (2013) MMLI: multimodal multiperson corpus of laughter in interaction. In: Salah AA, Hung H, Aran O, Gunes H (eds) HBU, Lecture notes in computer science, vol 8212. Springer, Hiedelberg, pp 184–195

87. Niewiadomski R, Mancini M, Ding Y, Pelachaud C, Volpe G (2014) Rhythmic body movements of laughter. In: Proceedings of the 16th international conference on multimodal interaction. ACM, New York, pp 299–306

88. O'Donnell Trujillo N, Adams K (1983) Heheh in conversation: some coordinating accomplishments of laughter. West J Commun (Includes communication reports) 47(2):175–191

89. Oh J, Wang G (2013) Laughter modulation: from speech to speech-laugh. In: Proceedings of the 14th annual conference of the international speech communication association (Interspeech). Lyon, France, pp 754–755

90. Oh J, Wang G (2013) Lolol: laugh out loud on laptop. In: Proceedings of the 2013 international conference on new musical instruments (NIME'13). Daejon, Korea

91. Owren M, Bachorowski JA (2003) Reconsidering the evolution of nonlinguistic communication: the case of laughter. J Nonverbal Behav 27(3):183–200

92. Pammi S, Khemiri H, Chollet G (2012) Laughter detection using alisp-based N-gram models. In: Proceeding of the interdisciplinary workshop on laughter and other non-verbal vocalisations. Dublin, Ireland, pp 16–17

93. Pecune F, Biancardi B, Ding Y, Pelachaud C, Mancini M, Varni G, Camurri A, Volpe G (2015) Lol-laugh out loud. In: Proceedings of AAAI 2015

94. Pelachaud C (2014) Interacting with socio-emotional agents. Procedia Comput Sci 39:4–7

95. Petridis S, Pantic M (2008) Fusion of audio and visual cues for laughter detection. In: International conference on content-based image and video retrieval, CIVR 2008. ACM, New York, pp 329–337. URL http://doc.utwente.nl/62669/

96. Petridis S, Pantic M (2011) Audiovisual discrimination between speech and laughter: why and when visual information might help. IEEE Trans Multimed 13(2):216–234. doi:10.1109/TMM.2010.2101586

97. Petridis S, Martinez B, Pantic M (2013) The mahnob laughter database. Image Vis Comput 31(2):186–202. doi:10.1016/j.imavis.2012.08.014
98. Piot B, Pietquin O, Geist M (2014) Predicting when to laugh with structured classification. In: Annual conference of the international speech communication association (Interspeech)
99. Platt T, Hofmann J, Ruch W, Niewiadomski R, Urbain J (2012) Experimental standards in research on AI and humor when considering psychology. In: Proceedings of fall symposium on artificial intelligence of humor
100. Platt T, Hofmann J, Ruch W, Proyer RT (2013) Duchenne display responses towards sixteen enjoyable emotions: individual differences between no and fear of being laughed at. Motiv Emot 37(4):776–786
101. Preuschoft S, van Hooff JARAM (1997) The social function of "smile" and "laughter": variations across primate species and societies. Lawrence erlbaum associates, Mahweh, New Jersey, pp 171–189
102. Qu B, Pammi S, Niewiadomski R, Chollet G (2012) Estimation of faps and intensities of aus based on real-time face tracking. In: Proceedings of the 3rd symposium on facial analysis and animation, FAA '12. ACM, New York, pp 13:1–13:1. doi:10.1145/2491599.2491612
103. Reuderink B (2007) Fusion for audio-visual laughter detection (2007). URL http://essay.utwente.nl/714/
104. Riek L, Rabinowitch T, Chakrabarti B, Robinson, P (2009) Empathizing with robots: fellow feeling along the anthropomorphic spectrum. In: 3rd International conference on affective computing and intelligent interaction and workshops 2009. ACII 2009, pp 1–6. doi:10.1109/ACII.2009.5349423
105. Rienks R (2007) Meetings in smart environments. implications of progressing technology. PhD thesis, University of Twente. ISBN: 978-90-365-2533-6, Number of pages: 201
106. Rothbart MK (1973) Laughter in young children. Psychol Bull 80(3):247–256
107. Ruch W (1993) The handbook of emotions, chapter Exhilaration and humor, pp 605–616. Guilford Press, New York
108. Ruch W (2012) Towards a new structural model of the sense of humor: preliminary findings. In: Proceedings of fall symposium on artificial intelligence of humor
109. Ruch W, Ekman P (2001) Emotion, qualia and consciousness, chapter The expressive pattern of laughter. World Scientic Publishers, Tokyo, pp 426–443
110. Ruch W, Hofmann J (2012) A temperament approach to humor. Humor and health promotion, pp 79–113
111. Ruch W, Hofmann J, Platt T (2013) Investigating facial features of four types of laughter in historic illustrations. Eur J Humour Res 1(1):99–118
112. Ruch W, Hofmann J, Platt T, Proyer R (2013) The state-of-the art in gelotophobia research: a review and some theoretical extensions. Humor Int J Humor Res 27(1):23–45
113. Ruch WF, Platt T, Hofmann J, Niewiadomski R, Urbain J, Mancini M, Dupont S (2014) Gelotophobia and the challenges of implementing laughter into virtual agents interactions. Front Human Neurosci 8:928
114. Ruch W, Hofmann J, Platt T (2015) Individual differences in gelotophobia and responses to laughter-eliciting emotions. Personal Individ Differ 72:117–121
115. Sathya AT, Sudheer K, Yegnanarayana B (2013) Synthesis of laughter by modifying excitation characteristics. J Acous Soc Am 133:3072–3082
116. Schuller B, Steidl S, Batliner A, Vinciarelli A, Scherer KR, Ringeval F, Chetouani M, Weninger F, Eyben F, Marchi E, Mortillaro M, Salamin H, Polychroniou A, Valente F, Kim S (2013) The interspeech 2013 computational paralinguistics challenge: social signals, conflict, emotion, autism. In: Interspeech. ISCA, pp 148–152
117. Sestito M, Umiltà MA, De Paola G, Fortunati R, Raballo A, Leuci E, Maffei S, Tonna M, Amore M, Maggini C et al (2013) Facial reactions in response to dynamic emotional stimuli in different modalities in patients suffering from schizophrenia: a behavioral and emg study. Front Human Neurosci 7:368
118. Shahid S, Krahmer E, Swerts M, Melder W, Neerincx M (2009) You make me happy: using an adaptive affective interface to investigate the effect of social presence on positive emotion

induction. In: 3rd International conference on affective computing and intelligent interaction and workshops 2009. ACII 2009, pp 1–6. doi:10.1109/ACII.2009.5349355

119. Sneddon I, McRorie M, McKeown G, Hanratty J (2012) The belfast induced natural emotion database. IEEE Trans Affect Comput 3(1):32–41. doi:10.1109/T-AFFC.2011.26

120. Sundaram S, Narayanan S (2007) Automatic acoustic synthesis of human-like laughter. J Acous Soc Am 121(1):527–535

121. Szameitat DP, Darwin CJ, Wildgruber D, Alter K, Szameitat AJ (2011) Acoustic correlates of emotional dimensions in laughter: arousal, dominance, and valence. Cognit Emot 25(4):599–611

122. Tanaka H, Campbell N (2014) Classification of social laughter in natural conversational speech. Comput Speech Lang 28(1):314–325

123. Tokuda K, Yoshimura T, Masuko T, Kobayashi T, Kitamura T (2000) Speech parameter generation algorithms for hmm-based speech synthesis. In: Proceedings of the IEEE international conference on acoustics, speech, and signal processing (ICASSP), vol 3. IEEE, New York, pp 1315–1318

124. Truong KP, van Leeuwen DA (2007) Automatic discrimination between laughter and speech. Speech Commun 49(2):144–158. doi:10.1016/j.specom.2007.01.001, http://www.sciencedirect.com/science/article/pii/S0167639307000027

125. Urbain J (2014) Acoustic laughter processing. PhD thesis, University of Mons

126. Urbain J, Dutoit T (2012) Measuring instantaneous laughter intensity from acoustic features. In: Proceeding of the interdisciplinary workshop on laughter and other non-verbal vocalisations. Dublin, Ireland, pp 18–19

127. Urbain J, Niewiadomski R, Bevacqua E, Dutoit T, Moinet A, Pelachaud C, Picart B, Tilmanne J, Wagner J (2010) Avlaughtercycle. J Multimodal User Interfaces 4(1):47–58. doi:10.1007/s12193-010-0053-1

128. Urbain J, Cakmak H, Dutoit T (2012) Development of HMM-based acoustic laughter synthesis. In: Interdisciplinary workshop on laughter and other non-verbal vocalisations in speech, Dublin, Ireland, pp 26–27

129. Urbain J, Niewiadomski R, Hofmann J, Bantegnie E, Baur T, Berthouze N, Cakmak H, Cruz R, Dupont S, Geist M, Griffin H, Lingenfelser F, Mancini M, Miranda M, McKeown G, Pammi S, Pietquin O, Piot B, Platt T, Ruch W, adn Volpe G, Wagner J (2012) Laugh machine. In: Proceedings of Enterface12. The 8th international summer workshop on multimodal interfaces

130. Urbain J, Çakmak H, Dutoit T (2013) Automatic phonetic transcription of laughter and its application to laughter synthesis. In: Proceedings of the 5th biannual humaine association conference on affective computing and intellignet interaction (ACII). Geneva, Switzerland, pp 153–158

131. Urbain J, Çakmak H, Dutoit T (2013) Evaluation of HMM-based laughter synthesis. In: Proceedings of the IEEE international conference on acoustics, speech, and signal processing (ICASSP), Vancouver, Canada, pp 7835–7839

132. Urbain J, Niewiadomski R, Mancini M, Griffin H, Çakmak H, Ach L, Volpe G (2013) Multimodal analysis of laughter for an interactive system. In: Proceedings of the INTETAIN 2013

133. Vinciarelli A, Pantic M, Heylen D, Pelachaud C, Poggi I, D'Errico F, Schroeder M (2012) Bridging the gap between social animal and unsocial machine: a survey of social signal processing. IEEE Trans Affect Comput 3(1):69–87. doi:10.1109/T-AFFC.2011.27

134. Urbain J, Çakmak H, Charlier A, Denti M, Dutoit T, Dupont S (2014) Arousal-driven synthesis of laughter. IEEE J Select Top Signal Process 8:273–284. doi:10.1109/JSTSP.2014.2309435

135. Wagner J, Lingenfelser F, André E (2013) Using phonetic patterns for detecting social cues in natural conversations. In: Bimbot F, Cerisara C, Fougeron C, Gravier G, Lamel L, Pellegrino F, Perrier P (eds) INTERSPEECH 2013, 14th Annual conference of the international speech communication association, Lyon, France, August 25–29. ISCA, pp 168–172

136. Wagner J, Lingenfelser F, Baur T, Damian I, Kistler F, André E (2013) The social signal interpretation (SSI) framework: multimodal signal processing and recognition in real-time.

Proceedings of the 21st ACM international conference on multimedia, MM '13. ACM, New York, pp 831–834
137. Yoshimura T, Tokuda K, Masuko T, Kobayashi T, Kitamura T (1999) Simultaneous modeling of spectrum, pitch and duration in HMM-based speech synthesis. In: Proceedings of Eurospeech. Budapest, Hungary

Chapter 10
Prosody Enhances Cognitive Infocommunication: Materials from the HuComTech Corpus

Laszlo Hunyadi, István Szekrényes and Hermina Kiss

Abstract The multimodal HuComTech corpus aims at annotating, studying and publishing data related to a wide spectrum of markers of human behavior in human-human spoken dialogues. By doing so the final goal is to both understand human cognitive behavior in conversational settings and contribute to the enhancement of human-machine interaction systems. One of the main issues still leaving wide spaces for further development is related to speech prosody, the understanding of its association with possible cognitive processes for the expression of emotions as well as the online production of speech utterances. Since the latter often results in incomplete structures, the study of the relation between grammatical incompleteness and prosody can both contribute to a better understanding of human cognition and the enhancement of cognitive infocommunication systems. The data and analyses presented in this paper are intended to serve both these purposes. Two different approaches will be presented as methods of data exploration: the study of static temporal alignments within the *ELAN* annotation tool, and the discovery of dynamic temporal patterns using the *Theme* framework.

Keywords Multimodal communication · Prodosy · Intonation · Spoken syntax · Cognitive processes

L. Hunyadi (✉) · I. Szekrényes · H. Kiss
Department of General and Applied Linguistics, University of Debrecen,
Egyetem tér 1, Debrecen 4032, Hungary
e-mail: hunyadi@unideb.hu

I. Szekrényes
e-mail: szekrenyes.istvan@arts.unideb.hu

H. Kiss
e-mail: kissh3@gmail.com

© Springer International Publishing Switzerland 2016   183
A. Esposito and L.C. Jain (eds.), *Toward Robotic Socially Believable Behaving Systems - Volume I*, Intelligent Systems Reference Library 105,
DOI 10.1007/978-3-319-31056-5_10

10.1 Introduction

In our verbally centered world we may often forget about the fact that the default, lexical meaning of a word or a phrase often differs from its intended meaning in a conversation. Indeed, we can use the word 'yes' for denial, or the word 'no' for assertion, all depending on how we produce these words. We can pronounce 'yes' with a high falling tone for assertion and a rising tone for doubt or even denial; we usually assign the meaning 'denial' for 'no' with a high falling tone and 'doubt' or 'assertion' with a rising tone. It does not, however, mean that lexical meaning as such is non existent: it only means that speech prosody, including intonation, intensity and duration are an essential building block of the information structure of an utterance. We can look at prosody as operating on the default lexical meaning of a word or a phrase with the effect of accommodating this meaning to the intensional context of the conversation. This function of accommodation is not necessarily secondary to the lexical meaning: it can go so far as the verbal content of the utterance may even be hidden and prosody alone prevails. Prosody is by no chance the entrance of a newborn baby into the spoken world: even without being able to use content words and words in general, it can deliver and reflect to the most essential pieces of information by simply selecting the elements of prosody suitable to the actual purpose of communication.

If prosody has this degree of importance in communicating information, it appears reasonable to believe that it needs to have an outstanding role in cognitive infocommunication as well. Significant advances in this respect have been made in text-to-speech systems to be used in an ever increasing number of applications [1–4]. With all these achievements, infocommunication will especially become more 'cognitive' if it takes an even more proactive route by considering the emotions and, through them, the intentions of the speaker(s). By employing a dynamically changing scheme of prosody suitable to the emotional-intentional content and context of the conversation even an interaction between a human and a non-human will become smoother, more flexible and, ultimately, more successful.

Even with these insights intended to improve human-machine interaction, one needs to keep in mind also some understandably fundamental differences between a human and a machine: we are not going to build machines by "recreating" or "reinventing" humans embodied in a machine, and it is not even an expectation on the human side: all we, humans, want in the long run is to build "cognitive machines" (cf. the Human Speechome Project [5]) with capabilities of interacting with humans at many but not necessarily all levels of human communication. These machines will essentially be based on the principle of multimodality ensuring that the interplay of the temporal and sequential alignment of markers of various modalities results in a complex of information that reaches beyond the mere sum of the individual pieces of information delivered by the individual modalities, without, however, being a copy of human behavior. Accordingly, these machines will embody "cognitive states" of their own but which are, at the same time, compatible with human cognition. This is the line of thought present in the current research in the emerging field of cognitive infocommunication [6, 7], and this is how the term "speechability" is to be understood: machines are expected to have capabilities both analogous to in certain ways and different in others from human speech [8]. As such, the description and study of the prosody of human speech relevant to cognitive infocommunication is to be both inclusive (extended to capture a wide spectrum of its physical and contextual properties including its relation to syntax, semantics and pragmatics) and restricted to those which are implementable for the given machine [9].

The multimodal HuComTech corpus of Hungarian dialogues is intended for the collection and supply of large amounts of data for learning more about the interplay between the verbal, lexical characteristics of conversations on the one hand, and its nonverbal, emotional and interpretative content on the other. In what follows, with focus on prosody, in particular intonation (pitch movement) we are going to present data on and investigate the interplay between intonation and events pointing to certain cognitive functions, such as emotions, incomplete syntactic constructions and utterance initial adverbials.

10.2   Material

The multimodal HuComTech corpus is a set of 50 h of dialogues comprising of 110 formal conversations (simulated job interviews) and 110 informal ones (guided by a standard scheme). Annotation work on the corpus started in 2009 and was completed by the end of 2015. Annotations include video annotations for gaze, head movement, hand movement, posture, facial expressions, audio annotations for transcription, fluency of speech, turn management, emotions, as well as prosody (the latter done automatically) for pitch movement, intensity and pause. In addition, the manual annotations are uniquely extended to spoken syntax and, to our knowledge also as first of its kind, to unimodal (video only) pragmatics complementing multimodal pragmatic annotation. All these different layers of annotation are meant to be studied simultaneously, allowing for the description of the eventual temporal and structural alignments of all available multimodal markers. Whereas our annotation data published and accessible in The Language Archive are available in the .eaf format of ELAN [1] they can also be accessed as an SQL compatible relational database.

For the purpose of this paper, we are going to look in more detail at the attributes of *(a) emotions* as represented in the category *facial expressions, (b) syntactic completeness/incompleteness* in the category *syntactic, (c) conversational adverbials* in the category *transcription*. In all cases we are going to analyze data referring to the interviewee only. Since we are going to present comparative data across several layers, and since data analysis is still ongoing, we will rely on data from 61 files (30 formal and 31 informal conversations, about 15 h of recording) only. Data will be cumulative, not distinguishing between the two types of conversation.

---

[1] https://hdl.handle.net/1839/00-0000-0000-001A-E17C-1@view

10.2.1   Facial Expressions

The following attributes were annotated all manually: *natural, happy, sad, surprised, recall, tense*. Only the video was used, the audio was not present in this work. In assigning these interpretative attributes the annotators (two for each file) followed the FACS facial action coding scheme [10] and inter-annotator agreement was ensured by regular consultations. In addition, before starting a new file each annotator observed the behavior of the participants on the recording for a reasonable time in order to "learn" the behavior of the given participant to be annotated.

The distribution of the above emotion attributes for the 61 files is as follows (total # of emotions annotations: 5816) in Table 10.1.

10.2.2   Syntactic Completeness/Incompleteness

The syntactic annotation has a particular importance in understanding the cognitive aspects of communication as organized via spoken utterances: since speech is in some way or another the online reflection of some underlying cognitive processes, and since these processes are certainly faster than speech production involving the physical movement of the participating articulatory muscles, it is often the case that the speaker will "fall short of time" to produce a fully organized—grammatical—sequence of words. This condition may eventually result in grammatically incomplete clauses or whole sentences that may be hard to understand without their context or may even be filtered out (left unanalyzed) in the flow of speech. With all these misgivings, such incomplete structures are part of everyday conversation, and as such need to be considered in designing a more complete human-machine interaction system [11].

The distribution of the syntactic attributes is as follows (total # of clauses: 16957) in Table 10.2.

(The rest of the syntactic types also annotated but not mentioned here reflect such missing elements as the verb, the grammatical subject, the object, the adverb, etc., which do not occur at either left or right clause boundaries and do not affect the overall prosody of a clause, therefore they are not considered here.)

Table 10.1   Distribution of emotion attributes

Natural	Happy	Sad	Surprise	Recall	Tense
929 (15.97%)	3802 (63.37%)	39 (0.67%)	246 (4.23%)	425 (7.31%)	375 (6.45%)

Table 10.2   Distribution of syntactic attributes

Type 1 (complete clause)	Type 13 (incomplete clause)	Type 14 (single sentence words)
2708 (15.97%)	729 (4.3%)	3130 (18.46%)

10.2.3   Prosody: Pitch Movement (Intonation)

The annotation of pitch movement has the following challenge: although the acoustic features of speech can be measured physically, intonation is the product of our perception. Whereas speech sounds as represented in a waveform are a complex of a wide range of frequencies and associated intensity values changing in minute units of time, we perceive them as a combination of four main parameters: fundamental frequency (F0), timbre, intensity, and duration, all with a much "smoother" temporal resolution. Out of these parameters it is fundamental frequency that contributes to our interpretation of the speech sound as intonation or speech melody. Thus we can see that the annotation of speech prosody is similar to, say, the annotation of the visual characteristics of nonverbal behavior: its formal aspects can be captured by certain physically definable parameters, but in a holistic sense these parameters are just the form, they actually are subjects to interpretation. And this is where one meets the challenge: how can we capture those data that are significant for our perception and disregard those which are not. For the HuComTech project an annotation tool, *ProsoTool* was developed [12, 20] to produce the automatic prosody annotation of the corpus with the aim to model human perception. As for pitch annotation, five levels of fundamental frequency were assigned to the pitch range of the given speaker, namely, from the deepest tone space to the highest tone space, as L_2, L_1, M, H_1, H_2. Pitch movements are assigned to well definable minimal tone contours and are characterized by the following relative attributes: *stagnant* (pitch remaining within the given tonal space), *descending* (lowering tone crossing one or more tonal spaces), *fall* (lowering tone rapidly crossing one or more tonal spaces), *upward* (rising tone crossing one or more tonal spaces), *rise* (rising tone rapidly crossing one or more tonal spaces). Accordingly, a falling tone can end up at a higher pitch than a rising tone, or a stagnant tone can be in a lower or higher tonal space than a rise or a fall. This interpretative annotation is supplemented by the annotation of the exact bordering pitch data, i.e. data reflecting the measured starting and ending frequency values of the given minimal tone contour. The overall tone contour of a longer speech segment (such as a clause or sentence) is the result of the concatenation of the component minimal tone contours, each referred to with one of the above labels.

The subcorpus we are describing here has the distribution of minimal tone contours is shown in Table 10.3.

The following is shown in two-member sequences are frequent sequences (see Table 10.4: total number of occurrences in the formal and the informal dialogues) (see Tables 10.5, 10.6, 10.7, 10.8).

Table 10.3 Distribution of tone contours

Fall	Rise	Stagnant	Descending	Upward
8173 (43.78%)	6435 (34.47%)	2350 (12.59%)	1018 (5.45%)	694 (3.72%)

Table 10.4 Two-member sequences

Sequence	Formal	Informal
Rise + fall	870	2012
Rise + rise	96	205
Fall + fall	212	437
Fall + rise	972	2119

Table 10.5 Sequence: rise + fall

	Formal		Informal	
	Rise	Fall	Rise	Fall
Average duration	709.8	862.4	657.5	777.9
Standard deviation	517.6	716.8	407.5	532

Table 10.6 Sequence: rise + rise

	Formal		Informal	
	Rise	Rise	Rise	Rise
Average duration	666.8	682.2	623.1	611.3
Standard deviation	517.4	478.6	377.6	354.8

Table 10.7 Sequence: fall + fall

	Formal		Informal	
	Fall	Fall	Fall	Fall
Average duration	746.9	808.8	700.5	768.8
Standard deviation	497.1	671.6	474.1	554.3

The following three-member sequences are frequent sequences (see Tables 10.9, 10.10, 10.11, 10.12, 10.13).

There are two relatively frequent sequences (see Table 10.14).

The above and similar shorter sequences of patterns may be repeated so as to be concatenated into longer stretches of a tonal contour. These longer stretches may, in principle, offer some more insight into the specific prosody of certain communicative events. From among the above patterns the following are repeated to form longer pieces of a tonal contour: *rise + fall* (formal: 132, informal: 445), *fall + rise* (formal: 161, informal: 463), *fall + stagnant + rise* (formal: 1, informal: 6), *rise + stagnant + fall* (formal: 1, informal: 1), *rise + rise + fall* (formal: 0, informal: 1), *fall + fall + rise* (formal: 1, informal: 2), *fall + stagnant + rise + fall* (formal: 0, informal:1).

Table 10.8 Sequence: fall + rise

	Formal		Informal	
	Fall	Rise	Fall	Rise
Average duration	846.7	691.7	801.6	675.4
Standard deviation	630.2	508.1	540.8	404.8

Table 10.9 Three-member sequences

Sequence	Formal	Informal
Fall + stagnant + rise	86	191
Rise + stagnant + fall	43	87
Rise + rise + fall	47	89
Fall + fall + rise	84	192

Table 10.10 Sequence: fall + stagnant + rise

	Formal			Informal		
	Fall	Stagnant	Rise	Fall	Stagnant	Rise
Average duration	574.9	1615.2	516.6	612.9	1350.7	516.4
Standard deviation	344.5	1531.9	195.9	329.8	1027.5	232.5

Table 10.11 Sequence: rise + stagnant + fall

	Formal			Informal		
	Rise	Stagnant	Fall	Rise	Stagnant	Fall
Average duration	413.1	1008.5	687.7	470.4	1002	542.6
Standard deviation	145.2	570.8	609.6	149.7	655.5	262.8

Clearly, it is the tonal segments with alternating directions (*fall* versus *rise*) that are the ones that appear to form part of longer tonal contours (Tables 10.15 and 10.16).

As a first approximation to associate one of the important attributes of prosody, i.e. pitch movement with any communicative functions, we tested the variance of the average duration of occurrences of the individual pitch patterns above in each of the 30 formal and 31 informal dialogues using F-test and repeated measures for the effect of the given dialogue types on the character of pitch movement. We found significant difference in the between-subject comparison, i.e. between the two dialogue types in all patterns (see in Table 10.17).

Table 10.12 Sequence: rise + rise + fall

	Formal			Informal		
	Rise	Rise	Fall	Rise	Rise	Fall
Average duration	612	679.7	774	613.7	657.2	772.7
Standard deviation	440.7	458.2	532	313.6	400.2	488.8

Table 10.13 Sequence: fall + fall + rise

	Formal			Informal		
	Fall	Fall	Rise	Fall	Fall	Rise
Average duration	765.9	823.3	623.5	749.7	731.8	632
Standard deviation	503.8	666.1	311.9	542.4	489.6	362.2

Table 10.14 Four-member sequences

Sequence	Formal	Informal
Fall + stagnant + rise + fall	38	98
Rise + stagnant + fall + rise	20	48

Table 10.15 Sequence: fall + stagnant + rise + fall

	Formal				Informal			
	Fall	Stagnant	Rise	Fall	Fall	Stagnant	Rise	Fall
Average duration	554.7	1508.8	504.3	799.4	611	1279.8	514.5	773.6
Standard deviation	230.2	1308.3	170.4	771	310.8	870.1	240.9	567.1

These data support our intuition and everyday experience: not only do speakers differ in the phonetics of their speech (voice), they also differ in the more global patterns of the way they speak, i.e. in their prosody. It is the within-subject variability of data that could shed light on the possible role of prosody (in our case: the formation of tonal contours). As for this latter, there were only two patterns found with a significant F-test in the within-subject comparison: *rise-fall* ($F = 4.2036, N = 29, p = 0.0495$), *rise-stagnant-fall-rise* ($F = 11.2435, N = 29, p = 0.0022$).

These data clearly suggest that the type of dialogue (whether it is conducted in a formal or informal manner) does indeed affect the way participants in the communicative event make use of speech prosody. It however further investigation is needed

Table 10.16 Sequence: rise + stagnant + fall + rise

	Formal				Informal			
	Rise	Stagnant	Fall	Rise	Rise	Stagnant	Fall	Rise
Average duration	370.7	1169.7	807	622.2	468.2	1040.9	580.8	618.4
Standard deviation	230.2	1308.3	170.4	771	310.8	870.1	240.9	567.1

Table 10.17 Between-subject comparison of pitch movements in formal and informal dialouges

Pattern	Result
Rise + fall	$F = 688.7305, N = 29, p > 0.0001$
Rise + rise	$F = 192.9638, N = 20, p > 0.0001$
Fall + fall	$F = 350.5749, N - 28, p > 0.0001$
Fall + rise	$F = 509.9192, N = 29, p > 0.0001$
Fall + stagnant + rise	$F = 259.8849, N = 22, p > 0.0001$
Rise + stagnant + fall	$F = 268.8140, N = 13, p > 0.0001$
Rise + rise + fall	$F = 340.0877, N = 14, p > 0.0001$
Fall + fall + rise	$F = 172.0493, N = 20, p > 0.0001$
Fall + stagnant + rise + fall	$F = 323.4265, N = 20, p > 0.0001$
Rise + stagnant + fall + rise	$F = 37.3226, N = 29, p > 0.0001$

to find out why these patterns involving an initial *rise* and including a *fall* appear to be those contours that show this specific divide between the formal and informal dialogues.

10.3   Methods

The fundamental challenge in dealing with multimodal behavioral data is that multimodality at a given point of time or for a given pattern of behavior cannot simply be captured by the complex of all data available at the given moment across all relevant modalities. It is an essential property of multimodality in behavior that this complex is dynamic: the availability of certain participating modalities may have both an intra- and inter-subject variability both across time and across event types. Accordingly, but probably strikingly at first glance, virtually any of the multimodal markers for virtually any behavior is optional. It has the consequence that one can hardly attribute a default set of modality markers to any single moment leading up to an event type. As an example, when someone shows agreement, one can say "Yes" accompanied by a nod and a gaze directed towards the interlocutor, however, none of these moments in any of these modality is obligatory: one can agree without saying "Yes" and/or nodding or gazing. Even more, one can even show/suggest agreement while saying "No" instead of "Yes", all this interpretation depending on yet another modality, prosody. Rather than considering such alignments of multimodal markers at single moments, our interpretation and judgment is based on evaluating the temporal variation of markers in various modalities, which, if repeated several times, will be considered as formal configurations constructed along the timeline. Those configurations which are interpreted as contributing to the expression of certain behavioral functions will be considered as patterns of behavior, whereas other configurations not associated with such functions will not be perceived as such. This understanding of the difference data the formal and the functional properties of configurations of signals underlines the role of interpretation for them to possibly become patterns of behavior. And, interpretation being also influenced by non-formal (contextual, attitudinal etc.) factors, the variable nature of behavioral "markers" (variable across subjects, both across time and event types) can obviously be justified. Therefore, in order to properly describe multimodality in behavior, one needs to consider the temporal alignments of various signals, the temporal configurations of such repeated signals and, finally, functionally evaluate these configurations as patterns of behavior.

There are several tools available for the study of the alignment of multimodal markers. In the next section we will present alignment data (and data on their linear sequence) using the software tool *ELAN* enabling us to find temporal alignments of markers within a preset threshold across an unlimited number of modalities and associate them with respective frequency data (for a description of the temporal sequences of multimodal data in the HuComTech corpus cf. [18]). By doing so we obtain a large amount of descriptive data offering a general view of the landscape of the composition of data by sequences and frequencies. The alignments and their sequences alone, however, do not tell us whether they form actual patterns characteristic of given behavioral patterns. Namely, we can find out whether our assumed alignments or their sequences are at all manifested in the data and, if they are, with what frequencies, but with these descriptions we cannot determine whether they represent real functional patterns and, even if they do, what is the probability that they are specific to a given behavior.

Due to the essential optionality of multimodal markers mentioned above we need to expect that patterns are at least partly hidden from the naked eye: they are composed of a potentially discontinuous sequence of markers, such that the distance between the elements considered adjacent in the pattern can be variable. Search for preset sequences as performed in *ELAN* or a similar environment will not result in such, typically behavioral patterns.

In order to identify at least some of the patterns underlying our behavioral data we use a radically different approach, offered by the software *Theme* [13]. Designed for the discovery of hidden patterns of behavior, this statistical framework searches for temporal patterns between events associated with time. The search is dynamic in the sense that it tests the statistical probability of the temporal configuration of virtually all events across all modalities and within a variable temporal distance between the events. Using our earlier example, it attempts to discover the temporal pattern (T-pattern) of the behavioral event type of "agreement" even if "Yes" is not immediately aligned with a nod or a given gaze. It is not looking for the frequency of a multimodal complex consisting of "Yes", a nod and a gaze, but it attempts to find what multimodal markers (events) are temporally associated with the behavioral function of "agreement" and with what probability, regardless of any other events possibly happening within the same time interval.

In the next section we will present data derived from the use of both methodologies mentioned: descriptive data about alignments and their sequences using *ELAN*, and interpretive data about possible temporal patterns using *Theme*. For the purpose of building a system of infocommunication, both kinds can prove to be useful: whereas the descriptive data contribute to building an inventory of the possible means of behavior, the interpretive ones contribute to understanding the multimodal nature of the given behavioral types.

10.4 Discussion

In this section it will be shown how the overall picture of individual attributes of facial expression (emotions), syntactic completeness/incompleteness and utterance-initial conversational adverbials are aligned with intonation (description using *ELAN*) and what configurations can functionally be discovered as actual behavioral patterns (probability and interpretation using *Theme*).

10.4.1 *Facial Expressions*

It can be observed that, in general, the expression of facial emotions coincides with or has some temporal precedence to its vocal (prosodic) expression. In the annotation of the HuComTech corpus we annotated perceived emotions at three different levels: at the level of video only (the annotators did not here the actual speech), audio only (the annotators had no access to the video) and video+audio (with simultaneous access to both modalities). In what follows, we will look at the video+audio condition and cases when the annotation of the facial emotion is shortly (<500 ms) followed by some pitch movement; cf. Table 10.18.

According to Table 10.18, the three emotions *natural, happy* and *recall* are the most frequent ones, indirectly reflecting the topics and mood of the conversations.

Table 10.18 Facial emotions followed by pitch movement

Emotion	Pitch contour					
	Formal/informal	Fall# (%) [average duration, ms]	Rise# (%) [average duration, ms]	Stagnant# (%) [average duration, ms]	Descending# (%) [average duration, ms]	Upward# (%) [average duration, ms]
Natural	F	88 (44.7) [794.7]	46 (42.6) [994.1]	17 (29.3) 1688.4	14 (58.3) [2300.3]	5 (31.3) [3357]
	I	122 (37.5) [860]	69 (34.8) [713.2]	34 (39.5) [1720.2]	17 (42.5) [2370.3]	12 (33.3) [1325.5]
Happy	F	28 (14.2) [724.9]	25 (23.1) [730.76]	8 (13.8) [2105/9]	3 (11.5) [6290.3]	2 (12.5) [3474.5]
	I	98 (30.2) [764.1]	66 (33.3) [647.9]	27 (31.4) [1189.6]	15 (37.5) [1714.2]	14 (38.9) [729.4]
Sad	F	3 (1.5) [469.3]	0	0	0	0
	I	3 (1.5) [741.5]	3 (1.5) [489.7]	0	1 (2.5) [2813]	1 (2.8) [4845]
Surprised	F	2 (1) [412.5]	2 (1.9) [604.5]	1 (1.7) [1350]	0	0
	I	2 (0.6) [1087.5]	5. (2.5) [811.2]	2 (2.3) [1321.5]	0	1 (2.8) [703]
Recall	F	60 (30.5) [777]	25 (23.1) [734.7]	22 (37.9) 1416.8	7 (26.9) [1820]	8 (50) [3097.4]
	I	92 (28.3) [725.4]	55 (27.8) [684.7]	23 (26.7) 1336.7	7 (17.5) [3809.3]	8 (22.2) [1752.6]
Tense	F	16 (8.1) [958.3]	10 (9.3) [617.9]	10 (17.2) [1007.7]	2 (7.7) [644.5]	1 (6.3) [806]
	I	0	0	0	0	0

Regarding the distribution of the total occurrences of the individual tonal contours, in the between-subject comparison we find that the distinction between formal and informal as dialogue types is significant for *natural* ($F = 1.0081, N = 9, p = 0.0242$) and *recall* ($F = 6.0973, N = 9, p = 0.0356$) only. *Tense* could not be tested due to the total lack of data in the informal dialogue. Regarding the average durations of the individual pitch contour types, the between-subject comparison shows a significant distinction between the two dialogue types in all types with sufficient data (*sad* and *tense* were not sufficiently manifested); cf. *natural*: $F = 7.3150, N = 9, p = 0.0242$, *happy*: $F = 9.2406, N = 4, p = 0.0384$, *surprise*: $F = 21.1170, N = 2, p = 0.0442$, *recall*: $F = 14.0392, N = 4, p = 0.0200$). The within-subject comparison was not significant in the absolute occurrence or average duration condition, either suggesting that the distribution by occurrence and duration of pitch contours, respectively, followed a fairly similar pattern across pitch contour types within the given dialogue types. In all, these statistical data tend to support our intuition that the type a given dialogue takes place in does actually influence the tonal structure of our speech. However, one can also argue that it is not just the frequency or duration of a given tonal contour that matters, but the concatenation of these contours into longer stretches of prosody that may become meaningful and discriminative for some communicative function. But in order to identify such patterns we must be aware that frequency and duration do not independently bear any function, instead, they are part of a more complex prosodic phenomenon: the alignment or sequence of prosodic constituents must be organized in a statistically significant way so that a given functional interpretation can be associated with them.

Thus in order to go beyond searching for alignments and their sequences on a predefined basis we used *Theme* in an attempt to automatically identify statistically significant patterns with behavioral function.

There are at least three important advantages of the approach offered by *Theme* in comparison to traditional descriptive statistics: first, one does not necessarily have to predefine those parameters and variables between and across which a pattern is to be identified, second, the interval between events potentially qualifying as constituents of a pattern does not need to be predefined either, third, and probably most importantly, the validity of patterns that are thus discovered are supported by statistical significance.

The corpus used in *Theme* was the same as the one used above for descriptive statistics including 30 formal dialogues of approximately 10 min of duration each and 31 informal dialogues with double that duration.

In the descriptive analysis shown in Table 10.18 we set the distance between two possible constituents of patterns to $<= 500$ ms. This intuition was essentially confirmed by *Theme*: within a user-defined interval of 2000 ms it set the *critical interval* (*Theme*'s term for such a possible distance) automatically to a range very close to the one used in our descriptive analysis: for the formal dialogues it was 589.9 ms (*stdev* = 307.2) and for the informal 725.7 (*stdev* = 303.5). The t-test showed that this difference was not significant (*Prob* > t : 0.5008, *correlation* : −0.034, N = 30).

According to the t-test, the formal and informal dialogues significantly differed in a number of variables: *number of event types in patterns* (formal: 12.6, *stdev* = 5.5, informal: 17, *stdev* = 5.1, *Prob* > *t* :< .0001*, *correlation* : 0.46709, *N* = 30), *number of different patterns* (formal: 51.9, *stdev* = 41.3, informal: 99.9, *stdev* = 76.1, *Prob* > *t* :< .0013, *correlation* : 0.28855, *N* = 30), *number of pattern occurrences* (formal: 1196.6, *stdev* = 1431.6, informal: 3185.5, *stdev* = 3367.8, *Prob* > *t* :< .0025, *correlation* : −0.2237, *N* = 30). Even though the informal dialogues were longer than the formal ones affecting the number of overall pattern occurrences, the higher number of different pattern types in the informal dialogues was essentially due to differences in the dialogue type: informal dialogues allow for more pattern variation than formal ones.

The patterns identified with $p <= 0.000005$, showed that patterns quite often select immediate constituents from the same modality; i.e. either some pitch movement selecting another pitch movement, or an emotion selecting another emotion (normally the beginning and end of a pitch or emotion, respectively, forming immediate constituents of a pattern); for some more complex unimodal patterns cf. the following examples (*b* and *e* are event types indicating the beginning and end of an associated event, respectively):

(*b*, *rise*(*e*, *rise*   *b*, *fall*))

$$ID = 466 \quad N = 1181 \quad Length = 3 \quad Duration = 1057614 \quad \%Duration = 6$$
$$Log10(P(1)) = -1, 26 \quad Log10(P(N)) = -323, 31$$

$$(10.1)$$

(*b*, *happy*, *reduced*(*e*, *natural*, *moderate*   *e*, *happy*, *reduced*))

$$ID = 394 \quad N = 26 \quad Length = 3 \quad Duration = 40026 \quad \%Duration = 0$$
$$Log10(P(1)) = -6, 83 \quad Log10(P(N)) = -323, 31$$

$$(10.2)$$

The full list of multimodal patterns (including both pitch movement and emotion) identified by *Theme* at significae level $p \leq 0.000005$ in the formal dialogues is as follows:

((*b*, *recall*   *b*, *fall*)*e*, *fall*)

$$ID = 970 \quad N = 106 \quad Length = 3 \quad Duration = 173385 \quad \%Duration = 1$$
$$Log10(P(1)) = -1, 16 \quad Log10(P(N)) = -18, 36$$

$$(10.3)$$

((*b*, *natural*   *b*, *fall*)*e*, *fall*)

$$ID = 886 \quad N = 105 \quad Length = 3 \quad Duration = 120081 \, \%Duration = 1$$
$$Log10(P(1)) = -1, 30 \quad Log10(P(N)) = -18, 79$$

$$(10.4)$$

(*b*, *fall*(*e*, *fall*   *e*, *recall*))

$$ID = 358 \quad N = 80 \quad Length = 3 \quad Duration = 97455 \quad \%Duration = 1$$
$$Log10(P(1)) = -2, 35 \quad Log10(P(N)) = -323, 31$$

$$(10.5)$$

$(b, rise(e, rise \; e, natural))$

$$ID = 470 \quad N = 71 \quad Length = 3 \quad Duration = 84265 \quad \%Duration = 1$$
$$Log10(P(1)) = -2, 31 \quad Log10(P(N)) = -323, 31$$
$$\text{(10.6)}$$

$(b, stagnant(e, stagnant \; e, recall))$

$$ID = 499 \quad N = 39 \quad Length = 3 \quad Duration = 73723 \quad \%Duration = 0$$
$$Log10(P(1)) = -2, 72 \quad Log10(P(N)) = -323, 31$$
$$\text{(10.7)}$$

Interestingly, we do not find the emotion *happy* in these patterns. Instead, the patterns include the emotions *recall* and *natural* only. The beginning of *recall* is associated with a beginning *falling* and its end with the end of a stagnant contour, both tonal contours corresponding to our most probable intuition: when one is faced with a question and is recalling, one starts with a falling pitch contour as a sign of searching in the memory, and, continues with a low (and *stagnant*) pitch contour until an answer is found.

Probably to some surprise, this was the only pattern found for informal dialogues, even though, as our descriptive statistics above showed, informal dialogues clearly contain many instances of emotions, including *happy*. Even if we may expect to find more behavioral patterns by narrowing down the predefined *p*-value for searches, the fact that at our very restricted significance ($p \leq 0.000005$) the emotion *recall* is identified in association with the tonal contour *fall* fully corresponds to our intuition described above, too.

$((b, recall \; b, fall)e, fall)$

$$ID = 1225 \quad N = 129 \quad Length = 3 \quad Duration = 178969 \quad \%Duration = 1$$
$$Log10(P(1)) = -1, 25 \quad Log10(P(N)) = -17, 71$$
$$\text{(10.8)}$$

10.4.2 Syntactic Completeness/Incompleteness

It has often been observed that duration in terms of pauses in certain syntactic positions plays an important role in speech segmentation in general [14] and in face-to-face conversation [15]. Variation in duration, however, often proves to be secondary to variation in frequency [16, 17], therefore, in order to improve a human-machine interaction system, we need to put focus on the frequency aspect of prosody as well. In the next table we show what kinds of pitch movements occur at most 500 ms before the end of a given clause (*Type 1*: a clause with all obligatory grammatical connections present, *Type 13*: a clause with several obligatory grammatical connections missing, *Type 14*: sentence words) (Table 10.19).

Table 10.19 Pitch movement at end of clause

Syntactic type		Formal/informal	Fall# (%) [average duration, ms]	Rise# (%) [average duration, ms]	Stagnant# (%) [average duration, ms]	Descending# (%) [average duration, ms]	Upward# (%) [average duration, ms]
	Pitch contour						
Type 1		F	74 (37.8) [933.2]	66 (51.2) [650.4]	18 (60) 1730.8	14 (77.8) [4035.6]	6 (50) [3577.5]
		I	279 (44.7) [835.3]	297 (56.9) [693]	110 (65.5) [1666.5]	40 (58) [2404.65]	5 (15.2) [3357]
Type 13		F	11 (37.8) [506.5]	13 (10.1) [780.7]	2 (6.7) [890]	1 (5.6) [3175]	2 (16.7) [1336]
		I	32 (5.1)[718.3]	61 (11.7) [742]	12 (7.1) [1145.3]	6 (8.7) [1755.7]	6 (18.2) [2492.8]
Type 14		F	111 (37.8) [565.1]	50 (38.8) [688.8]	10 (33.3) [729.4]	3 (16.7) 1177	4 (33.3) [2698.5]
		I	313 (50.2) [583.7]	164 (31.4) [552.3]	46 (27.4) [662.3]	23 (33.3) [791.8]	22 (66.7) [792.5]

Regarding the number of the total occurrences of a given tonal contour the between-subject comparison does not show any statistically significant difference between the uses of the given syntactic types in the formal and the informal dialogue conditions. Regarding the average duration of the tonal contour types, however, the test reveals a significant difference between them (*Type 1*: $F = 11.7304, N = 4$, $p = 0.0267$, *Type 13*: $F = 14.3082, N = 4, p = 0.0194$, *Type 14*: $F = 17.8828$, $N = 4, p = 0.0134$). The within-subject comparison shows no significance either for total occurrence or duration. These data suggest that, indeed, the prosodic formation of the three syntactic types above (complete clauses with at least one grammatical connection, clauses with multiple missing grammatical connections and complete clauses consisting of single words) have significantly different prosodic realization based on the given (formal or informal) dialogue type. What exactly such realizations may mean these data do not reveal. For that, we would need to go beyond the single alignments and predefined sequences of pitch contours and clause types. In order to discover patterns consisting of longer stretches of prosodic contours and the associated clause types we should, paradoxically, need to know in advance the constituents of the patterns we are just trying to identify. Obviously, traditional descriptive/statistical methodologies are not suitable to do this job. What we may want to try next, instead, is to let *Theme*, the statistical environment designed for identifying possible hidden patterns of events organized along the timeline find out how syntactic types and pitch contours may form various, statistically significant patterns that may underlie the perceived functional difference between formal and informal types of dialogues.

As we saw before, for descriptive purposes we *ad hoc* predefined the pattern internal interval at <500 ms. Setting *Theme* to limit *critical interval* to 2000 ms, this intuition was essentially confirmed by *Theme* finding candidates for patterns with an average *critical interval* close to that value: 602.9 ms (*stdev* = 227.3. *ms*) for the formal, and 618.8 ms (*stdev* = 142.2) for the informal dialogues (the values for the two conditions significantly correlating: *Prob* > *t* : 0.3520, *correlations* : 0.02916, $N = 30$). According to the t-test, the formal and informal dialogues significantly differed in a number of variables: *number of event types in patterns* (formal: 19.2, *stdev* = 5.3, informal: 25.7, *stdev* = 2.9, *Prob* > *t* :< .0001*, *correlation* : 0.07656, $N = 30$), *number of different patterns* (formal: 63, *v* = 39.9, informal: 200.5, *stdev* = 98.6, *Prob* > *t* :< .0001*, *correlation* : 0.04469, $N = 30$), *number of pattern occurrences* (formal: 842.9, *stdev* = 932.5, informal: 3532.7, *stdev* = 2875.3, *Prob* > *t* :< .0001*, *correlation* : −0.0867, $N = 30$).

The patterns identified with $p <= 0.000005$, showed a preference for immediate constituents from the same modality; i.e. either some pitch movement forming a pattern with another pitch movement, or a clause with another clause; cf. the following examples (*b* and *e* are event types indicating the beginning and end of an associated event, respectively):

$(b, rise(e, rise \quad b, stagnant))$

$ID = 335 \quad N = 332 \quad Length = 3 \quad Duration = 384402 \quad \%Duration = 2$

$Log10(P(1)) = -1,81 \quad Log10(P(N)) = -323,31$

(10.9)

(the beginning of a rising tone is followed by the end of a rising tone followed by the beginning of a stagnant tone, the latter following the end of the rise within an interval shorter than the duration of the rising tone itself)

$(b, synt14(e, synt14 \quad b, synt7))$

$ID = 457 \quad N = 29 \quad Length = 3 \quad Duration = 24740 \quad \%Duration = 0$

$Log10(P(1)) = -3,07 \quad Log10(P(N)) = -323,31$ (10.10)

(a *Type 14* clause is followed by a *Type 7* clause in an interval shorter than the duration of the *Type 14* clause itself)

The list of all patterns multimodal (including pitch movement and clause beginning/end) by frequency in *formal* dialogues is as follow:

$(b, rise(e, rise \quad e, synt14))$

$ID = 339 \quad N = 64 \quad Length = 3 \quad Duration = 51146 \quad \%Duration = 0$

$Log10(P(1)) = -2,51 \quad Log10(P(N)) = -323,31$ (10.11)

$(b, rise(e, risee, synt7))$

$ID = 340 \quad N = 58 \quad Length = 3 \quad Duration = 64736 \quad \%Duration = 0$

$Log10(P(1)) = -2,46 \quad Log10(P(N)) = -323,31$ (10.12)

$(b, rise(e, rise \quad b, synt9))$

$ID = 337 \quad N = 29 \quad Length = 3 \quad Duration = 22049 \quad \%Duration = 0$

$Log10(P(1)) = -2,91 \quad Log10(P(N)) = -18,66$

(10.13)

$(b, rise(e, riseb, synt13))$

$ID = 336 \quad N = 18 \quad Length = 3 \quad Duration = 18133 \quad \%Duration = 0$

$Log10(P(1)) = -3,22 \quad Log10(P(N)) = -14,46$

(10.14)

$(b, stagnant(e, stagnant \quad e, synt14))$

$ID = 374 \quad N = 18 \quad Length = 3 \quad Duration = 17207 \quad \%Duration = 0$

$Log10(P(1)) = -3,27 \quad Log10(P(N)) = -323,31$

(10.15)

These data confirm that the end of clauses of *Type 14* (complete, one word clauses) can form a pattern with some pitch contour further specifying what contours are part of that pattern (the end of *rise* or *stagnant*). *Type 13* (incomplete) clauses are also confirmed to be associated with some pitch contour, but, contrary to our intuition reflected in the descriptive data, this association is at the beginning rather than the end of the clause. *Type 1* (complete) clauses were not found to be associated with a specific tonal contour, most probably due to the high variability of the duration of such clauses. Unexpectedly, however, *Theme* found patterns associated with *Type 7* (missing logical subject) and *Type 9* (missing object) including a pitch rise at the beginning and the end of the clause, respectively. We need to point out, however, that using *Theme* we did not study the pattern forming capacity of all possible concurrent variables, therefore, a study with a wider scope could lead to the discovery of further patterns at the crossroads of syntactic type and pitch movement.

The list of all multimodal patterns (including pitch movement and clause beginning/end) by frequency in *informal* dialogues is as follow:

$(b, synt15(e, synt15 \ \ b, rise))$

$$ID = 1138 \ \ N = 267 \ \ Length = 3 \ \ Duration = 326358 \ \ \%Duration = 1$$
$$Log10(P(1)) = -2,05 \ \ Log10(P(N)) = -323,31$$
$$(10.16)$$

$(b, rise(e, rise \ \ b, synt13))$

$$ID = 628 \ \ N = 193 \ \ Length = 3 \ \ Duration = 219526 \ \ \%Duration = 1$$
$$Log10(P(1)) = -2,21 \ \ Log10(P(N)) = -323,31$$
$$(10.17)$$

$((e, rise \ \ b, synt13)e, synt13)$

$$ID = 8226 \ \ N = 178 \ \ Length = 3 \ \ Duration = 254583 \ \ \%Duration = 1$$
$$Log10(P(1)) = -2,81 \ \ Log10(P(N)) = -323,31$$
$$(10.18)$$

$((b, synt14 \ \ b, rise)e, rise)$

$$ID = 5281 \ \ N = 151 \ \ Length = 3 \ \ Duration = 106395 \ \ \%Duration = 0$$
$$Log10(P(1)) = -2,12 \ \ Log10(P(N)) = -323,31$$
$$(10.19)$$

$((b, synt14 \ \ b, rise)e, synt14)$

$$ID = 5282 \ \ N = 145 \ \ Length = 3 \ \ Duration = 78596 \ \ \%Duration = 0$$
$$Log10(P(1)) = -2,23 \ \ Log10(P(N)) = -323,31$$
$$(10.20)$$

$(b, rise(e, rise \ \ e, synt14))$

$$ID = 631 \ \ N = 133 \ \ Length = 3 \ \ Duration = 87534 \ \ \%Duration = 0$$
$$Log10(P(1)) = -2,36 \ \ Log10(P(N)) = -323,31$$
$$(10.21)$$

$(b, synt9(e, synt9 \quad e, stagnant))$

$\quad ID = 1543 \quad N = 127 \quad Length = 3 \quad Duration = 238926 \quad \%Duration = 1$

$$Log10(P(1)) = -2, 32 \quad Log10(P(N)) = -323, 31$$

$$(10.22)$$

$(b, synt14(e, rise \quad e, synt14))$

$\quad ID = 1000 \quad N = 124 \quad Length = 3 \quad Duration = 88672 \quad \%Duration = 0$

$$Log10(P(1)) = -2, 31 \quad Log10(P(N)) = -323, 31$$

$$(10.23)$$

$(b, synt15(e, synt15 \quad b, stagnant))$

$\quad ID = 1139 \quad N = 111 \quad Length = 3 \quad Duration = 144595 \quad \%Duration = 0$

$$Log10(P(1)) = -2, 47 \quad Log10(P(N)) = -323, 31$$

$$(10.24)$$

$(b, synt15(e, synt15 \quad e, stagnant))$

$\quad ID = 1147 \quad N = 102 \quad Length = 3 \quad Duration = 134391 \quad \%Duration = 0$

$$Log10(P(1)) = -2, 63 \quad Log10(P(N)) = -323, 31$$

$$(10.25)$$

$((e, stagnant \quad b, synt13)e, synt13)$

$\quad ID = 9086 \quad N = 102 \quad Length = 3 \quad Duration = 161976 \quad \%Duration = 1$

$$Log10(P(1)) = -2, 68 \quad Log10(P(N)) = -323, 31$$

$$(10.26)$$

$(b, rise(e, rise \quad b, synt7))$

$\quad ID = 629 \quad N = 80 \quad Length = 3 \quad Duration = 86706 \quad \%Duration = 0$

$$Log10(P(1)) = -2, 66 \quad Log10(P(N)) = -323, 31$$

$$(10.27)$$

$(b, synt14(e, stagnant \quad e, synt14))$

$\quad ID = 1001 \quad N = 60 \quad Length = 3 \quad Duration = 39866 \quad \%Duration = 0$

$$Log10(P(1)) = -2, 74 \quad Log10(P(N)) = -323, 31$$

$$(10.28)$$

$(b, synt15(e, synt15 \quad e, descending))$

$\quad ID = 1146 \quad N = 50 \quad Length = 3 \quad Duration = 67356 \quad \%Duration = 0$

$$Log10(P(1)) = -2, 96 \quad Log10(P(N)) = -19, 27$$

$$(10.29)$$

$(b, rise(e, rise \quad b, synt10))$

$\quad ID = 627 \quad N = 22 \quad Length = 3 \quad Duration = 16668 \quad \%Duration = 0$

$$Log10(P(1)) = -3, 69 \quad Log10(P(N)) = -18, 96$$

$$(10.30)$$

When comparing the set of patterns found by *Theme* in the formal and the informal conditions, the first we notice is that the informal dialogues show a larger number and wider variety of patterns. A probable explanation is that it is the reflection of the larger variability of behavior in an informal context in general, both due to the variability of the context and the individual reactions to it. However, the fact that these are patterns occurring in large numbers across subjects demonstrates that, regardless of the various conditions underlying the variability of behavior there are still specific general patterns the individual subjects follow.

The informal condition shows again that these patterns are mostly associated with Type 14 clauses and at the end of the clause. *Type 13* clauses have tonal patterns both at the beginning and the end of the clause. We again find patterns with *Type 7* and *Type 9* clauses. There are 22 instances of patterns including *Type 10* clauses (missing an adverbial complement) and some pitch movement (*rise*). It is especially worth noting that *Type 15* of syntactic insertion is found with high frequency forming a prosodic pattern: either associated with the beginning of a *rise* or the end of a *descending* tone: this is the case when there is no grammatical means available for denoting a syntactic relation (that of insertion) and that it is prosody that assumes this marking function [19].

10.5   Conclusions

The multimodal HuComTech corpus with its rich annotation scheme ranging from descriptive and interpretative video and audio annotation to syntax, prosody and pragmatics offers an opportunity to attempt to find alignments and sequences of markers across a wide spectrum of modalities. The method presented in this paper for identifying dynamic and partly hidden temporal patterns beyond temporal alignments goes a significant step further in learning about the structural and functional relations embedded in behavior.

As a result of both descriptive and structural analyses we presented data underlying the possible role of prosody in enhancing human-machine interaction systems: the variation of pitch movement in relation to syntax and expressions of emotions can inherently play an important role in delivering information about the various cognitive states of the speaker in spoken dialogues. Since prosody, including our present topic, intonation, is strongly associated with our cognition, a more enhanced employment of prosody solutions in human-machine interaction will further enhance the development of the complex and growing scientific field of cognitive infocommunication as well.

Acknowledgments Research presented in this paper was partly supported by project TÁMOP 4.2.2-C/11/1/KONV-2012 and OTKA NK116402. The work was made possible by the generous support of NeDiMAH (Network for Digital Methods in the Arts and Humanities), a cross-European project of the European Science Foundation.

References

1. Sagisaka Y, Campbell N, Higuchi N (eds) (1996) Computing prosody: computational models for processing spontaneous speech. Springer, New York
2. Rajeswari KC, Uma Maheswari P (2012) Prosody modeling techniques for text-to-speech synthesis systems—A survey. Int J Comput Appl (0975–8887) 39(16):8
3. Teixeira JP (2012) Prosody generation model for TTS systems: segmental durations and F0 contours with fujisaki model. LAP LAMBERT Academic Publishing
4. Chaloupka Z, Hork P (2012) Prosody modelling for TTS systems using statistical methods. In: Cognitive behavioural systems, COST 2102 International training school, Dresden, Germany, February 21–26, 2011. Revised Selected Papers, Springer, Heidelberg, pp 174–183
5. Roy BC, Frank MC, Roy D (2012) Relating activity contexts to early word learning in dense longitudinal data. In: Proceedings of the 34th annual meeting of the cognitive science society. Sapporo, 2012
6. Baranyi P, Csapo A (2012) Definition and synergies of cognitive infocommunications. Acta Polytech Hung 9(1):67–83
7. Sallai G (2012) Defining infocommunications and related terms. Acta Polytech Hung 9(6):5–15
8. Baranyi P, Csapo A, Varlaki P (2014) An overview of research trends in coginfocom. In: IEEE International conference on intelligent engineering systems, Tihany, pp 181–186
9. Hunyadi L (2011) Multimodal human-computer interaction technologies. Theoretical modeling and application in speech processing, Argumentum 7, pp 240–260
10. Ekman P, Friesen W (1978) Facial action coding system: a technique for the measurement of facial movement. Consulting Psychologists Press, Palo Alto
11. Hunyadi L, Incompleteness and fragmentation in spoken language syntax and its relation to prosody and gesturing: cognitive processes versus possible formal cues. Knowledge-based information systems in practice. Springer (to appear)
12. Szekrnyes I (2014) Annotation and interpretation of prosodic data in the HuComTech corpus for multimodal user interfaces. J Multimodal User Interfaces 8(2):143–150
13. Magnusson MS (1996) Hidden real-time patterns in intra- and inter-individual behavior: description and detection. Eur J Psychol Assess 12(2):112–123
14. Ladd DR (1996) Intonational phonology. Cambridge University Press, Cambridge
15. Edlund J, Heldner M, Hirschberg J (2009) Pause and gap length in face-to-face interaction. In: Proceedings of Interspeech 2009, Brighton
16. Hunyadi L (2010) Cognitive grouping and recursion in prosody. In: Hulst H van der (ed) Recursion and human language, de Guyter, Berlin & New York, pp 343–370
17. Hunyadi L (2002) Hungarian sentence prosody and universal grammar. Peter Lang, New York
18. Abuczki A (2011) A multimodal analysis of the sequential organization of verbal and nonverbal interaction. Argumentum 7:261–279
19. Hunyadi L (2010) Cognitive grouping and recursion in prosody. In: Hulst, Harry van der (ed) Recursion and human language. Studies in Generative Grammar [SGG] 104. de Gruyter Mouton, pp 343–370
20. Szekrenyes I (2015) ProsoTool, a method for automatic annotation of fundamental frequency. In: Cognitive Infocommunications (CogInfoCom), 2015 6th IEEE International Conference on, 19–21 Oct. 2015, Györ, IEEE 2015, pp 291–296

Chapter 11
Analysis of Emotional Speech—A Review

P. Gangamohan, Sudarsana Reddy Kadiri and B. Yegnanarayana

Abstract Speech carries information not only about the lexical content, but also about the age, gender, signature and emotional state of the speaker. Speech in different emotional states is accompanied by distinct changes in the production mechanism. In this chapter, we present a review of analysis methods used for emotional speech. In particular, we focus on the issues in data collection, feature representations and development of automatic emotion recognition systems. The significance of the excitation source component of speech production in emotional states is examined in detail. The derived excitation source features are shown to carry the emotion correlates.

11.1 Introduction

Humans have evolved various forms of communication like facial expressions, gestures, body postures, speech, etc. The form of communication depends on the context of interaction, and is often accompanied by various physiological reactions such as changes in the heart rate, skin resistance, temperature, muscle activity and blood pressure. All forms of human communication carry information at two levels, the message and the underlying emotional state.

Emotions are essential part of real life communication among human beings. Various descriptions of the term emotion are studied in [21, 22, 60, 88, 92, 98, 100]. Some of the descriptions are:

(a) "Emotions are underlying states which are evolved and adaptive. Emotion expressions are produced by the communicative value of underlying states" [22].

P. Gangamohan (✉) · S.R. Kadiri · B. Yegnanarayana
International Institute of Information Technology, Hyderabad, India
e-mail: gangamohan.p@students.iiit.ac.in

S.R. Kadiri
e-mail: sudarsanareddy.kadiri@research.iiit.ac.in

B. Yegnanarayana
e-mail: yegna@iiit.ac.in

© Springer International Publishing Switzerland 2016
A. Esposito and L.C. Jain (eds.), *Toward Robotic Socially Believable Behaving Systems - Volume I*, Intelligent Systems Reference Library 105, DOI 10.1007/978-3-319-31056-5_11

(b) "Emotions are experienced when something unexpected happens at times," [92].

(c) "Emotions are an organism's interface to outside world," and carry three principle functions, *the significance and relevance of particular stimuli, preparation of the organism's physiology for appropriate action,* and *communication of the organism's state and intention to other organisms* [98].

Many studies reported the relationship between emotions and expressive states. Darwin [22] defined emotion as an inner state and expression as a communicative manifestation of the underlying emotional state. Studies in the literature [15, 88, 92, 99, 119] have referred to the concept of "basic emotions". Typically, "basic emotions" are perceived cross-cultural, while non-basic emotions are learnt in a culture-specific manner [15]. Although the concept of "basic emotions" has been widely accepted, there has been considerable debate on the composition and number of "basic emotions" [27, 115]. Typically, anger, happiness, fear, sadness and neutral are identified as "basic emotions".

In speech communication, humans have the natural ability to grasp the underlying emotional state, as well as the lexical content. A fundamental research problem is, given a speech signal, can the underlying emotional state of the speaker be identified? Normally, human beings perceive emotions of an unknown speaker through deviations from normal state. For example, in the perception of angry speech, there is increase in voice intensity, raise in pitch and faster speaking rate. From this, one can infer that there is reference (neutral/normal), and deviations from the reference are being perceived. The objective is to derive an emotion-specific feature representation, by focusing on the deviations in the components of the speech production mechanism.

Progress in human-computer voice interaction systems is limited due to difficulty of the machine in recognizing and responding to even basic emotions. Whereas, it happens effortlessly in human-human communication. Systems recognizing speaker's emotions, and also responding expressively, are essential for natural interaction. Research on emotions in speech has applications in spoken dialogue systems, automated response systems, call centers, etc.

This chapter gives a review of emotional speech analysis, along with description of studies addressing specific research issues in emotion analysis. The organization of the chapter is as follows. The data collection issues for emotion related studies are discussed in Sect. 11.2. A review of studies on emotion analysis and emotion recognition is given in Sect. 11.3. Studies addressing some specific issues are presented in Sect. 11.4. Finally, Sect. 11.5 discusses some research challenges in emotion analysis.

11.2 Data Collection of Emotional Speech

Progress in applications related to emotional speech relies heavily on the availability of suitable databases [44]. Emotion databases developed by different research groups can be categorized into simulated, semi-natural and natural databases [26,

73, 112, 126]. Simulated parallel emotion databases are collected from speakers (voice-talents) by prompting them to enact emotions through specified text in a given language. The simulated parallel emotion databases in [17, 28, 74, 113, 138] were collected from speakers by asking them to emote the same text in different emotions. The main disadvantage of such databases is that deliberately enacted emotions are quite at variance from 'spontaneous' emotions, and also at times they are out of context [30, 126]. Semi-natural is a kind of enacted corpus, where the context is given to the speakers. The semi-natural emotion database in German language was developed by asking speakers to enact the scripted scenarios eliciting each emotion [13, 116]. The third kind of emotion database is a natural database, where the recordings do not involve any prompting or eliciting of emotional responses. Sources for such natural situations could be talk shows, interviews, panel discussions and group interactions in TV broadcast. A brief overview of databases is given in Table 11.1.

For developing high quality expressive text-to-speech (TTS) synthesis systems, a large natural database of each target emotion is required [103]. But it is impractical to obtain a large natural emotion database. Emotion conversion systems are adopted as a post-processing block for speech synthesis systems. In this, a large database of neutral speech is used to generate speech by a TTS system, which is then fed to an emotion conversion system [88, 103]. The output speech from these systems is unnatural, and also has the constraint of requiring parallel speech corpus [31].

As the collection of natural databases is mostly carried out from TV talk shows, call centers and interaction with robots, speakers involved in these cases control their emotion/expressive states. There is a trade-off between the controllability and naturalness of the interaction [51]. There is also difficulty in identifying the emotion or expressive state of a dialogue. Therefore, the annotation is described mostly by three basic primitives or dimensions, namely, valence, arousal/activation and dominance/power [51, 125]. Also, there are two other cases of ambiguity in annotating. One of them is occurrence of mixed emotions in an utterance. For example, there is a possibility of combinations like, surprise-happy, frustration-anger and anger-sad, occurring in the dialogue. The second reason for ambiguity is due to unsustainabililty of emotion throughout the dialogue. In natural communication among human beings, emotion may not be sustainable over the entire duration of the dialogue. The emotion is expressed mostly in some segments of the dialogue, with the rest of the dialogue being neutral. It is also difficult to define the boundaries for the occurrence of an emotion in continuous speech [64].

11.3 Emotional Speech Analysis

Feature representation plays a key role in developing any emotion related applications. The features used in many analysis studies can be broadly categorized into prosody, voice quality and spectral features. In this Section, the overview of these features is given along with the literature of emotion recognition studies.

Table 11.1 An overview of emotional speech data collection

	Simulated	Semi-natural	Natural
Description	Collected from trained speakers by asking them to emote same text in different emotions	Developed by asking speakers to enact the scripted scenarios eliciting each emotion	• Recording does not involve any prompting or the obvious elicitation of emotional responses
			• Sources: TV broadcast, call center calls, court rooms, etc.
Examples	• LDC [138]	• IEMOCAP [18]	• VAM [51]
	• EMO-DB [17]	• Belfast [116]	• Call centers [77, 87]
	• DES [28]	• NIMITEK [44]	• AIBO [117]
	• IITKGP-SESC [74]		
Advantages	• Most commonly used and standardized	• Near to natural database	• Completely natural
	• Comparison of results is easy	• Even though contextual information is available, it is still artificial	• Useful for real world emotion systems modeling
	• Number of emotions available are large		
Disadvantages	• It tells how emotions should be portrayed rather than how they are portrayed	• Less number of emotions are available	• Emotions are continuous
	• Context, environment and purpose dependent information is absent	• If speakers are aware that they are being recorded, the emotions become unnatural/artificial	• All emotions are not available
	• Episodic in nature, not true in real-world situations		• Contains multiple and concurrent emotions
			• Difficult to model
			• Copyright and privacy issues arise
Preferred applications	• Emotion conversion systems	• Emotion recognition systems	• Emotion recognition systems
	• Expressive speech synthesis systems	• Dimensional emotion analysis	• Dimensional emotion analysis
		• Categorical emotion analysis	• Categorical emotion analysis

Table 11.2 Summary of prosody and spectral features of emotional speech w.r.t. neutral speech

	Angry	Happy	Sad	Fear
Prosody features				
Average F_0	Higher	Higher	Lower	Much higher
F_0 range	Wider	Wider	–	Wider
F_0 contour (shape)	Irregular fluctuations	Descending and ascending patterns at irregular intervals	Downward inflections	Much irregular up-down fluctuations
Average intensity	Higher	Higher	Lower	Normal
Intensity range	Wider	Wider	Narrower	Normal
Speaking rate	Slightly faster	Normal	Slightly slower	Faster
Spectral features				
F_1 mean	Increase	–	Increase	Increase
F_1 bandwidth	–	Increase	Decrease	Decrease
F_2 mean	Increase	–	Decrease	–
High frequency energy	Increase	Increase	Decrease	Increase

11.3.1 Prosody Features

Efforts have been made to understand the contribution of speech prosody features towards production of emotion. Early studies were based on the fundamental frequency (F_0) of emotional speech [23, 36, 37, 40, 132, 133]. Along with F_0, various aspects of prosody like speaking rate, relative durations and intensity were examined [24, 50, 87]. One of the early studies using speaking rate parameter was by [36], where the speaking rate was described in terms of words per minute. The analysis of pauses in angry, happy, fear and sad speech was also performed in the study. The related features of speaking rate like the average duration of voiced speech, ratio of voiced to unvoiced speech, average duration of pauses and syllables per second, have been used in [7, 23, 24, 50, 58, 87]. A summary of the prosody features corresponding to basic emotions is given in Table 11.2. Some of the features share similar properties across different emotions. As observed from Table 11.2, angry and happy utterances have similar trends in F_0 and speaking rate, with respect to (w.r.t.) neutral speech.

11.3.2 Voice Quality Features

In a broader sense, the term voice quality refers to the characteristic auditory colouring of an individual's speech [67]. In general, speakers have their own voice quality signature. By varying voice qualities, they convey important information like intentions, emotions and attitudes. From the perspective of Laver's approach [75], voice quality is expressed in terms of laryngeal and supralaryngeal settings. Laryngeal settings are described by phonation types, pitch and loudness ranges. Supralaryngeal settings are described by longitudinal, latitudinal, tension modifications of vocal tract, and nasalisation.

In the literature, many studies performed voice quality analysis by considering the laryngeal activity (mainly phonations). Non modal phonations are often observed in emotional speech. Breathiness is associated with angry and happy speech [71, 88]. Vocal fry voice is observed in sad and relaxed speech [48, 71], which may be because of very low fundamental frequencies in these cases. Harsh voice, which corresponds to irregularity in voicing, was observed in fear speech [71].

Studies [4, 20, 46, 47] have shown that glottal source parameters like closed quotient (CQ), abruptness in closing and normalized amplitude quotient (NAQ) are useful in distinguishing different phonations. Also, these glottal source parameters are analyzed for emotional speech [1, 121, 122, 131]. These features were extracted from the glottal waveform derived using inverse filtering (IF) technique [3, 96]. There are several limitations in IF based approaches such as deriving the accurate transfer function by canceling out the effect of the vocal tract system, and obtaining the closed phase duration of the glottal cycle [38, 45]. Although these glottal source parameters give emotion correlates, the dynamic ranges of these features are observed to be speaker-specific [1, 48].

11.3.3 Spectral Features

The spectrum characterized by formant frequencies and their respective bandwidths is extensively analyzed for emotional speech [13, 100, 133]. In [133], it is observed that the vowels in angry speech are produced with wide open vocal tract, and inferred that the first formant (F_1) has higher mean than that of neutral speech. The predictions of formants (F_1, F_2 and F_3), their bandwidths and high frequency energy for the emotion classes are made in [13], which are given in Table 11.2 along with prosody features.

It is also interesting to note that there are certain changes in the spectral component which are associated with the glottal source excitation [54, 86, 133]. The syllables produced with higher fundamental frequencies in angry speech tend to have weaker F_1 amplitudes [133]. More closed phase of glottis configuration results in relatively higher amplitudes at high frequencies [86].

A fundamental characteristic of the spectrum of a speech signal is that it is sound-specific [72, 83]. The deviations in spectral features are analyzed for utterances having the same lexical content [87, 100, 133]. Also, some studies on spectral features [53, 78, 128] have shown that changes in the magnitude and shift of the formants in emotional states vary across vowels.

11.3.4 Emotions as Points in Continuous Dimensional Space

Viewing emotions as points in continuous dimensional space was first suggested in [101]. Emotions are mainly viewed as combinations of three dimensions/primitives, namely, valence, arousal/activion and dominance/power [101, 102, 104]. Several studies [12, 23, 93, 105] have explored the relation of features like F_0, durations, loudness and spectral parameters, to these dimensions. Some important findings of these studies are that high arousal speech (like angry and happy) is associated with increase in the average F_0, wider F_0 range and decrease in spectral tilt. Low arousal speech (like sad and disgust) is associated with decrease in the average F_0 and narrow F_0 range.

11.3.5 Studies on Emotion Recognition

Feature representation is the most important step for developing an automatic emotion recognition system. The prosody-related features are statistical measures of F_0 and intensity contours, and features related to speaking rate [29, 77, 80, 82, 107, 137]. The spectral features are mel frequency cepstral coefficients (MFCCs), linear prediction cepstral coefficients (LPCCs), modulation spectral features, formant frequencies, bandwidths of formant frequencies [55, 61, 77, 107, 134]. The voice quality features are shimmer, jitter and NAQ, which are related to the glottal excitation characteristics [32, 34, 49, 95, 97, 120, 121, 123].

Feature representation is done at different levels of speech such as frame level [56, 59, 68, 80, 81, 91], segment level [25, 61, 84, 90] and utterance level [19, 29, 55, 76, 82, 85, 97, 127, 137]. A common approach adopted for feature representation at segment and utterance levels is by statistical analysis of frame level features and prosody features. Some of the toolkits which are widely used for feature extraction are PRAAT (prosody and voice quality features) [16], APARAT (voice quality features) [2], OpenSMILE (prosody, voice quality and spectral features) [35] and OpenEAR (prosody and spectral features) [33].

There are two major approaches used in the automatic emotion recognition task. In the first approach, the utterances in a corpus are labeled with discrete emotion categories such as anger, happiness, sadness and fear. Feature representations at different levels are used for training models like (GMMs) [29, 61, 69, 81], support vector machines (SVMs) [55, 97, 137], artificial neural networks (ANNs) [6], deep neural networks (DNNs) [5], hidden Markov models (HMMs) [59, 80, 91]. The second approach is a primitive-based classification, where emotions are described by arousal, valence and dominance [32, 39, 50, 76, 82, 123], and the classification is done using a hierarchical binary tree approach. For example, the activation states are detected initially, and then the identification of valence states follows.

Most of the pattern recognition algorithms used for developing emotion recognition systems require large amount of labeled emotion data. Also, emotion recognition systems that use spectral feature representations require a phonetically/phonemically balanced database. In [32, 82], it is reported that prosody features can effectively discriminate activation states, but there is difficulty in discriminating valence states.

There are a few emotion recognition systems developed using linguistic features [14, 25, 29, 68, 77, 94, 97, 106, 108–111, 118]. Typically there are two approaches, one is a language modeling approach [25, 29, 68, 77, 94] and the other is a bag-of-words approach [14, 97, 106, 108, 111, 118]. In language modeling approach, utterances are transcribed manually, then language models are developed for each emotion. In the testing phase, the decision of emotion is given by the likelihood scores of each model for the test utterance. The bag-of-words approach was primarily developed for document retrieval tasks [63, 106]. In this approach each word in the vocabulary adds a dimension to the linguistic vector. The linguistic vector which gives the word term frequency within the utterance is used for classification of emotions. The limitations of these approaches are that the data required for each emotion should be large, and also these methods are based on ASR framework with its own limitations.

11.3.6 Limitations of the Studies

The prosody and voice quality features are mostly speaker-specific, and the spectral features are mostly sound-specific. In order to use these features for developing emotion recognition systems, a phonetically/phonemically balanced database covering several speakers might be required.

It is also observed that there are many interrelations among the type of database used, features, approaches and evaluation procedures. Some of the emotion recognition studies with various features, databases, pattern recognition algorithms and recognition accuracies are given in Table 11.3.

The following are some important observations from Table 11.3.

- The performance in terms of accuracy of emotion recognition systems built and tested using simulated parallel corpus [50, 61, 82, 120, 137] is high when compared to systems built and tested with semi-natural or natural databases [11, 32, 62, 76, 97]. As per spectral analysis of emotions [78, 87, 100, 128], for speech segments of the same lexical content, there exist deviations in the formant frequencies, bandwidths and spectral tilt, in somewhat emotion-specific way. These spectral deviations might help in discriminating emotions in the case of simulated parallel corpus.
- There are a few emotion recognition studies focused on cross-corpora and cross-lingual aspects [34, 62, 113, 120]. There are two ways of cross-corpora evaluations, a system developed using a database and tested with other database, or a mixture of cross-corpora used for both training and testing systems [113, 114, 120]. The reported recognition results are low in these cases.
- In several studies [6, 50, 61, 62, 69, 80, 82, 127], it is reported that there is confusion between anger and happiness emotions. The anger versus happiness confusion is also observed in the case of temporal modeling of intonation variations using HMMs [80].

Table 11.3 Review of some emotion recognition studies

References	Features	Database and no. of emotions	Pattern recognition algorithms	Accuracy and remarks
Jeon et al. [61]	Spectral and prosody	EMO-DB (simulated parallel), 4 emotions (anger, happiness, sadness and neutral)	GMM	85 % and more anger versus happiness confusion
Yeh and Chi [137]	Spectral and prosody	EMO-DB (simulated parallel), 7 emotions (anger, happiness, fear, disgust, boredom, sadness and neutral)	SVM	83 % and confusion matrix not reported
Sun and Moore II [120]	Spectral, prosody and voice quality	EPST, EMA and EMO-DB (all are simulated parallel), 4 emotions (anger, happiness, sadness and neutral)	SVM	Training and testing with same database, EPST—64 %, EMA—80 % and EMO-DB—83 %. Accuracy < 50 % for cross-corpora
Lugger and Yang [82]	Prosody and voice quality	EMO-DB (simulated parallel), 7 emotions	GMM	66 % and more anger versus happiness confusion
Grimm et al. [50]	Spectral and prosody	EMA (simulated parallel), 4 emotions (anger, happiness, sadness and neutral)	kNN(k-nearest neighbors)	83 % and more anger versus happiness confusion
Espinosa et al. [32]	Spectral, prosody and voice quality	VAM (natural), 3 categories (arousal, valence and dominance)	SVM	68 %
Rozgic et al. [97]	Spectral, prosody and lexical	USC-IEMOCAP (semi-natural), 4 emotions (anger, happiness, sadness and neutral)	SVM	69 %
Lee et al. [76]	Spectral, prosody and voice quality	AIBO (natural), 5 emotions (anger, emphatic, positive, neutral and others)	Bayesian logic regression	49 %

(continued)

Table 11.3 (continued)

References	Features	Database and no. of emotions	Pattern recognition algorithms	Accuracy and remarks
Atassi and Esposito [10]	Prosody and voice quality (emotion selective)	EMO-DB (simulated parallel) 6 emotions (anger, happiness, fear, disgust, boredom and sadness)	GMM	80 % and more anger versus happiness confusion
Ververidis and Kotropoulos [127]	Spectral and prosody	DES (simulated parallel), 5 emotions (anger, surprise, sadness, happiness and neutral)	GMM	More anger versus happiness confusion
Wu et al. [134]	Modulation spectral	VAM (natural), 3 categories (arousal, valence and dominance). EMO-DB (simulated parallel) 7 emotions	SVM	VAM—67 %, EMO-DB—85 %
Lin et al. [80]	Spectral and prosody	MHMC (semi-natural)	HMM	79 % and more anger versus happiness confusion
Atassi et al. [11]	Spectral, prosody and voice quality	COST 2102 Italian (natural) database, 6 emotions (anger, fear, happiness, irony, sadness and surprise)	ANNs	60.7 %

- There are a few studies on emotion recognition which addressed speaker dependent and speaker independent criteria [50, 70, 82, 85]. In these studies, the recognition accuracies of speaker independent systems are reported to be low when compared to the accuracies of speaker dependent systems. This may be because of the speaker-specific variations in the voice quality and prosodic features.

11.4 Studies on Some Specific Research Issues

The following are the specific research issues addressed in some recent studies

- *Analysis-by-synthesis to explore relative contribution of different components of emotional speech*: These experiments are performed using a flexible analysis-synthesis tool (FAST) [41]. This tool is used to modify the components of speech of the source utterance to that of the target, and vice-versa. These analysis-by-synthesis experiments are discussed in Sect. 11.4.1.
- *Identification of emotion-specific regions of speech*: Speaker-specific neutral versus non neutral regions are detected from speech signals using neural networks [65], and the studies are described in Sect. 11.4.2.
- *Significance of excitation source features in emotional speech*: The excitation source features extracted around the glottal closure instants (GCIs) are analyzed [42]. These studies are described in Sect. 11.4.3. A speaker-specific emotion recognition system developed based on these features [66] is described in Sect. 11.4.4.
- *Anger and happiness discrimination*: Emotion recognition studies indicate that there exists confusion among higher activation states like anger and happiness. Features related to the excitation source of speech are examined for discriminating anger and happiness emotions [43], and the studies are described in Sect. 11.4.5.
- *Discrimination between high arousal and falsetto voices*: Speakers tend to increase pitch when they raise their voice intensity in natural communication. Speakers can also shift to a falsetto register, where the pitch can be deliberately increased without raising the voice intensity. The excitation source related features are analyzed for these two cases in Sect. 11.4.6.

11.4.1 Analysis-by-Synthesis to Explore Relative Contribution of Different Components of Emotional Speech

The objective of this study is to determine the components of speech that contribute to the perception of emotion. A controlled set of experiments are conducted to modify the speech signal of the same sentence in neutral and emotional states. This is accomplished by using a flexible analysis-synthesis tool (FAST) described in [41].

Analysis of the relative contribution of F_0 and the amplitude of the speech signal is given in [79]. Similarly, experiments investigating the role of F_0 contour and duration parameter are described in [129]. A time-domain pitch synchronous overlap and add (PSOLA) algorithm is used for modification of these components. The general observation is that the modification of parameters individually in neutral speech does not give the perception of emotion.

In addition to the prosody-related components, components highlighting the characteristics of the excitation source and the vocal tract system are also considered in this section. The following five components of speech are used: Vocal tract shape, excitation source, excitation energy, F_0 contour and relative durations. The time varying vocal tract information is represented by linear prediction coefficients (LPCs) for each frame of 20 ms with a frame shift of 10 ms. The excitation source information is represented by the linear prediction (LP) residual signal within each glottal cycle. The glottal cycle relates to a signal between two adjacent epochs (or GCIs). These epochs are extracted using the zero frequency filtering (ZFF) method [89]. In the zero frequency filtered signal, the negative to positive zero crossings corresponds to the GCIs. The F_0 contour is obtained from the intervals between epochs. The energy of the LP residual within each glottal cycle gives the excitation energy. The set of experiments (denoted as $E1$ to $E31$ in Table 11.4) are performed to modify the components of the source utterance to that of the target utterance.

The IIT-KGP SESC database in Telugu language [74] is used in this study. From this database, utterances of 5 emotions (neutral, anger, happiness, sadness and fear) of the same sentence of two speakers are considered. Two sets of studies are made: Set-1 consists of modification of neutral to emotion, and set-2 consists of modification of emotion to neutral. The modified speech is subjected for evaluation. The subjective evaluation is carried out by 10 (student) listeners from the Speech and Vision Laboratory at IIIT Hyderabad. Each subject was asked to give a similarity score ranging from 1 to 5 for a pair of utterances. The score 5 indicates that the utterances have high similarity. The score 1 indicates that both the utterances are very much different. The original target utterance and synthesized speech of each emotion category are used for determining the similarity.

The experiments with average similarity scores greater than 3.0 for different emotion categories in both the sets are given in Table 11.4. It is interesting to note that the entries are almost same in both the sets, indicating that the parameters for modifying neutral to emotion and vice-versa gives similar perception of target utterance. It is interesting to note that modification of any one component is not adequate for creating or suppressing the characteristics of any emotion category. Experiments which include modification of F_0, duration and LPCs components seem to be having high scores in the case of anger category. Angry speech is produced under extreme displeasure and frustration, it exhibits very high and wider F_0 when compared to other emotions [88]. Also, it is observed that there is wide opening of vocal tract during speech in anger state [133]. Therefore, combination of these parameters might give a better perception of anger. For the perception of sadness, happiness and fear emotions, several combination of the components are possible.

Table 11.4 The analysis-by-synthesis experiments with average similarity scores greater than 3.0 for two sets

Exp.	Components for modification	Set-1 (neutral to emotion)	Set-2 (emotion to neutral)
E1	F_0	–	–
E2	Duration (dur)	–	–
E3	LPCs	–	–
E4	Excitation (exc) energy	–	–
E5	Excitation (exc) source	–	–
E6	F_0 and dur	AN(3.2)	–
E7	F_0 and LPCs	HA(3.1)	HA(3.1)
E8	F_0 and exc energy	–	–
E9	F_0 and exc source	SA(3.1)	
E10	Dur and LPCs	–	–
E11	Dur and exc energy	–	–
E12	Dur and exc source	–	–
E13	LPCs and exc energy	–	–
E14	LPCs and exc source	–	–
E15	Exc energy and exc source	–	–
E16	F_0, dur and LPCs	AN(3.7), HA(3.7), SA(3.1), FE(3.7)	AN(3.5), HA(3.2), SA(3.1), FE(3.6)
E17	F_0, dur and exc energy	FE(3.6)	FE(3.2)
E18	F_0, dur and exc source	SA(3.0), FE(3.5)	SA(3.1)
E19	F_0, exc energy and LPCs	HA(3.9)	
E20	F_0, exc source and LPCs	HA(3.0), SA(3.7)	SA(3.6)
E21	F_0, exc energy and exc source		SA(3.0)
E22	Dur, exc energy and LPCs	–	–
E23	Dur, exc source and LPCs	–	–
E24	Dur, exc energy and exc source	–	–
E25	Exc energy, exc source and LPCs	–	–
E26	F_0, dur, exc energy and exc source	AN(3.9), HA(3.4), SA(3.4), FE(3.7)	AN(3.9), HA(3.5), SA(3.4), FE(3.6)
E27	F_0, dur, exc energy and LPCs	AN(4.3), HA(3.8), SA(3.4), FE(3.9)	AN(4.0), HA(3.6), SA(3.5), FE(3.5)
E28	F_0, dur, exc source and LPCs	AN(4.1), HA(3.3), SA(3.6), FE(3.7)	AN(3.7), HA(3.5), SA(3.7), FE(3.6)
E29	F_0, exc energy, exc source and LPCs	SA(3.7)	HA(3.3), SA(3.8)
E30	Dur, exc energy, exc source and LPCs –	–	
E31	F_0, Dur, exc energy, exc source and LPCs	AN(4.4), HA(4.5), SA(4.3), FE(4.2)	AN(4.4), HA(4.1), SA(4.3), FE(4.5)

(*AN*—anger, *HA*—happiness, *SA*—sadness, and *FE*—fear)

Table 11.5 Neutral versus non neutral discrimination for EMO-DB and IIIT-H Telugu emotion databases [65]

	EMO-DB database (%)	IIIT-H Telugu database (%)
Excitation source	91.03	94.01
Vocal tract system	83.25	88.23

11.4.2 Identification of Emotion-Specific Regions of Speech—Neutral Versus Non Neutral Speech

In this section, the issue of (speaker-specific) neutral versus non neutral speech detection is discussed using models of neutral speech as reference. Autoassociative neural network (AANN) models are developed for capturing the excitation source and the vocal tract system components separately. For this purpose, LP residual and LPCs are used as approximations of the excitation source and vocal tract system components, respectively. A 10th order LP analysis with a frame length of approximately twice the instantaneous pitch period of the speech signal is chosen. For extracting the excitation source information, a 4 ms segment of the LP residual around each GCI is chosen. The vocal tract system characteristics are represented by the 15 dimensional weighted LPCC vector derived from the LPCs.

The network structure $33L\ 80N\ xN\ 80N\ 33L$ is chosen for developing the model for the excitation source component of the neutral speech. Here L refers to linear units, N refers to nonlinear ($tanh()$) output function of units, and x refers to the number of units in the compression layer. The structure $15L\ 40N\ xN\ 40N\ 15L$ is used for developing the model for the vocal tract system component of neutral speech. For both the AANN models, a universal background model (UBM) is developed using 15 s of neutral speech from each speaker. Speaker-specific AANN models for neutral speech are developed by training over the UBM using approximately 20 s of neutral speech data from a speaker. In the testing phase, the emotional speech utterance is presented to the neutral speech AANN models, and the mean squared error between the output and input is normalized with the magnitude of the input. Since the AANN models are developed using neutral speech, it is expected that the error should provide discrimination between neutral and emotional speech. It is observed that, error values are high when the test utterance is not neutral, and low when the test utterance is neutral. Using a threshold on the averaged normalized error value, the emotion regions can be detected.

The results of neutral versus non neutral detection using the ANN models for EMO-DB (in German language) and IIIT-H Telugu emotion database [42] are shown in the Table 11.5. From the Table 11.5, it appears that the excitation source information provides better discrimination of neutral versus non neutral states than the vocal tract system information. It is observed that the proposed excitation source and vocal tract system AANN models provide an improvement of approximately 10 and 3 %, respectively, over the recently proposed method [8, 9] (accuracy is 80.4 %)

for EMO-DB. It is also observed that the high arousal emotion states (like anger and happiness) are more discriminative compared to the low arousal emotion states (sadness and boredom). This is in conformity with the studies reported in [61, 76, 114]. It is to be noted that emotion information may not be uniformly distributed across all frames in time. It is also necessary to explore methods to combine the evidence from the excitation source and from the vocal tract system characteristics.

11.4.3 Significance of Excitation Source Features in Emotional Speech

This study demonstrates the significance of the excitation source features. The excitation features are extracted around the epoch locations. The following four features considered for this study: Instantaneous F_0, strength of excitation (SoE), energy of excitation (EoE) and loudness parameter (η). The instantaneous F_0 and SoE features are extracted using the ZFF method [89]. The slope of the zero frequency filtered signal at each epoch location is called SoE. The SoE parameter is related to the strength of the impulse-like excitation at the epoch [136]. The EoE feature is computed using the energy of the samples of the Hilbert envelope (HE) of the LP residual over 2 ms around each epoch. The loudness measure η gives the abruptness of the glottal closure [52]. The η feature is given by the ratio of the standard deviation and mean of the samples of the HE of the LP residual over 2 ms around each epoch location.

The deviations of the excitation source features of an emotional speech w.r.t. neutral speech are analyzed in the following six 2-dimensional (2-D) feature spaces: F_0 versus SoE, F_0 versus EoE, F_0 versus η, SoE versus EoE, η versus SoE and η versus EoE. Sample 2-D scatter plots for a pair of neutral and angry utterances are shown in Fig. 11.1.

The Kullback-Leibler (KL) distance [57] measure of the distributions in the 2-D feature spaces of neutral (reference) and emotion (test) utterances is used for representing the deviations. The KL distance measure is given by

$$D = \frac{1}{2}\left(tr(\Sigma_1^{-1}\Sigma_0) + (\mu_1 - \mu_0)^T \Sigma_1^{-1}(\mu_1 - \mu_0) - k - ln\left(\frac{det\,\Sigma_0}{det\,\Sigma_1}\right)\right), \quad (11.1)$$

where D is the KL distance, k is the dimension of the features space, and Σ_0, Σ_1 are the covariance matrices, and μ_0, μ_1 are the mean vectors of the neutral and emotion utterances, respectively.

Three databases, namely, IIIT-H Telugu, IITKGP-SESC and EMO-DB are used in this study. In the case of IIIT-H Telugu (semi-natural) database, 3 test utterances for each emotion (anger, happiness, sadness and neutral) and 2 neutral (reference) utterances for each speaker are used. The KL distance averaged over different test and reference utterances are given for two speakers (1 male and 1 female) in Table 11.6. From Table 11.6, the average KL distance values are low when the reference and test utterances are both neutral. The average KL distance values are higher when the

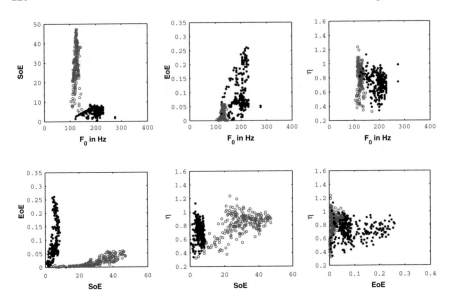

Fig. 11.1 2-D scatter plots of a neutral utterance (marked by 'o') and angry utterance (marked by '∗') of a speaker [42]

Table 11.6 Average KL distance values between reference (neutral (NU)) utterance and test (angry (AN), happy (HA), sad (SA) and NU) utterances, for IIIT-H Telugu emotion database [42]

	2-D feature spaces					
	F_0 versus SoE	F_0 versus SoE	F_0 versus η	SoE versus SoE	η versus SoE	η versus EoE
Speaker 1						
NU versus NU	0.02	0.02	0.01	0.07	0.02	0.02
NU versus AN	2.10	69.70	1.60	110.00	0.40	93.00
NU versus HA	0.89	28.29	0.70	44.35	0.17	27.56
NU versus SA	0.96	0.91	0.77	0.49	0.27	0.13
Speaker 2						
NU versus NU	0.06	0.09	0.01	0.17	0.06	0.09
NU versus AN	0.10	47.50	0.10	66.20	0.20	46.70
NU versus HA	0.41	22.09	0.25	312.74	0.15	20.90
NU versus SA	0.06	0.29	0.08	0.29	0.07	0.30

test utterance is not neutral, which indicates that the speakers modify the excitation characteristics while producing emotional speech.

In the above case, the reference and test utterances have different lexical content. In order to study the effectiveness of the excitation source features, the reference and test utterances with the same lexical content are considered for the simulated parallel databases (IITKGP-SESC and EMO-DB). The average KL distance values for these databases are given in Table 11.7. The average KL distance values are low when the

Table 11.7 Average KL distance values between reference (neutral (NU)) utterance and test (angry (AN), happy (HA), sad (SA) and NU) utterances, for IITKGP-SESC and EMO-DB databases [42]

| | IITKGP-SESC | | | | | | EMO-DB | | | | | |
| | 2-D feature spaces | | | | | | 2-D feature spaces | | | | | |
	F_0 versus SoE	F_0 versus EoE	F_0 versus η	SoE versus EoE	η versus SoE	η versus EoE	F_0 versus SoE	F_0 versus EoE	F_0 versus η	SoE versus EoE	η versus SoE	η versus EoE
Speaker 1												
NU versus NU	0.07	2.08	0.01	2.14	0.06	2.26	0.11	0.33	0.02	0.40	0.10	0.35
NU versus AN	0.29	86.86	0.04	86.76	0.26	98.58	2.19	4.57	1.01	3.70	0.70	4.01
NU versus HA	0.57	14.84	0.24	15.04	0.31	16.83	0.24	1.34	1.13	2.38	0.85	0.92
NU versus SA	0.28	1.01	0.08	1.14	0.18	1.10	2.33	2.59	0.13	4.05	1.53	2.51
Speaker 2												
NU versus NU	0.07	0.10	0.03	0.15	0.04	0.09	0.67	0.25	0.05	0.82	0.61	0.28
NU versus AN	0.35	38.20	0.13	53.20	0.21	38.60	2.41	18.13	1.47	17.66	0.85	19.68
NU versus HA	0.38	4.90	0.19	5.24	0.08	3.46	3.71	15.21	7.41	11.51	5.48	12.42
NU versus SA	0.21	0.55	0.18	1.68	0.06	1.48	5.72	17.90	13.51	19.10	6.92	40.02

reference and test utterances are both neutral, higher otherwise. This indicates that the excitation source features are independent of the lexical content.

Apart from the above general observations, the average KL distance values are observed to be dependent on speaker, emotion and even culture. Among all the considered emotion categories, these values are high in the case of anger. One of the reasons is due to higher and wider F_0 variations in angry speech [88]. As anger is extremely charged state, the resulting F_0 variations might be because of increased vocal effort. The values in the case of sadness for IIIT-H Telugu and IITKGP-SESC databases are observed to be low. In the case of EMO-DB database, the values are observed to be higher for sad speech.

11.4.4 Emotion Recognition System Based on Excitation Source Features

A system for automatic recognition of emotions is developed based on the excitation source feature deviations of emotional speech w.r.t. neutral speech. The emotions considered for this study are: anger, happiness, neutral and sadness. Three excitation source features, namely, instantaneous F_0, SoE and EoE are used. Distributions of pairs of F_0, SoE and EoE are used (see Fig. 11.1) to derive feature representations for the emotion recognition system [42].

The distributions of neutral utterances are first normalized as follows: Let the distributions of F_0, SoE and EoE for a neutral utterance be denoted by R_{F_0}, R_{SoE} and R_{EoE}, and for an emotion utterance by E_{F_0}, E_{SoE} and E_{EoE}, respectively. Let $R_{m_{F_0}}$, $R_{m_{SoE}}$ and $R_{m_{EoE}}$ represent the mean values, and $R_{\sigma_{F_0}}$, $R_{\sigma_{SoE}}$ and $R_{\sigma_{EoE}}$ represent the standard deviations of the distributions of R_{F_0}, R_{SoE} and R_{EoE}, respectively. The distributions are normalized w.r.t. mean and standard deviation. The normalized distribution for R_{F_0} is given by

$$N_{R_{F_0}} = \frac{R_{F_0} - R_{m_{F_0}}}{R_{\sigma_{F_0}}}. \tag{11.2}$$

Similarly, the normalized distributions $N_{R_{SoE}}$ and $N_{R_{EoE}}$ are obtained for R_{SoE} and R_{EoE}, respectively.

The distributions of the features of an emotion utterance are normalized w.r.t. the neutral utterance as follows. The normalized distribution of E_{F_0} is given by

$$N_{E_{F_0}} = \frac{E_{F_0} - R_{m_{F_0}}}{R_{\sigma_{F_0}}}. \tag{11.3}$$

Similarly, the normalized distributions $N_{E_{SoE}}$ and $N_{E_{EoE}}$ are obtained for E_{SoE} and E_{EoE}, respectively. The normalization is carried out in a speaker-specific way using the speaker's neutral utterance.

Three 2-D feature distributions, $D1$: ($N_{E_{F_0}}$ versus $N_{E_{SoE}}$), $D2$: ($N_{E_{EoE}}$ versus $N_{E_{F_0}}$) and $D3$: ($N_{E_{EoE}}$ versus $N_{E_{SoE}}$) are modeled by a Gaussian distribution, represented by mean vector and covariance matrix. These distributions capture the emotion-specific feature deviations. In the training phase, for each speaker, for each emotion utterance, the 2-D feature distributions (templates) are extracted. In the testing phase, a speaker's neutral speech is required to obtain the normalized distributions for the test utterances. The KL distance scores are computed among the corresponding 2-D distributions of test utterance and stored templates. The emotion category with maximum matched templates is declared for the test utterance.

A 2-stage binary hierarchical classification is implemented. In Stage 1, anger and happiness emotions are grouped into one class, and sadness and neutral emotions are grouped into another class. In Stage 2, the comparisons are made between neutral versus sadness categories, and anger versus happiness categories.

The confusion matrices for IIIT-H Telugu emotion database after Stage 1 and Stage 2 are shown in Tables 11.8 and 11.9, respectively. From the values listed in Table 11.8, the binary classification at stage 1 gives 96 % accuracy. This is in conformity with the studies [76], where generally the acoustic features effectively discriminate between high arousal and low arousal emotions. From Table 11.9, it is observed that the confusion between anger and happiness states is high. The recognition for neutral, sadness, anger and happiness emotions are 91.2, 97, 71.43 and 52 %, respectively, giving a total recognition accuracy of 79.23 % for the 4 class problem.

The proposed method is also applied on the EMO-DB database, and the results are shown in Table 11.10. The emotion recognition at Stage 2 of EMO-DB is 75 %. The performance of the EMO-DB is low because the confusions between angry and happy utterances are observed high.

The results of the proposed method indicate that the features corresponding to the excitation source seem to carry emotion-specific information. The performance of the system can be improved by increasing the number of reference (trained) templates and speakers.

11.4.5   Discrimination of Anger and Happiness

The production characteristics of angry and happy speech are examined to determine features that can discriminate these two emotions. In particular, the closed quotient (C_q) of the glottal vibration, SoE and the ratio of high to low frequency band energies are used.

Table 11.8 Confusion matrix after Stage 1 for IIIT-H Telugu emotion database [66]

	Neutral/sad	Angry/happy
Neutral/sad	68/68	0/68
Angry/happy	5/62	57/62

Table 11.9 Confusion matrix after Stage 2 for IIIT-H Telugu emotion database [66]

	Neutral	Sad	Angry	Happy
Neutral	31/34	3/34	0/34	0/34
Sad	1/34	33/34	0/34	0/34
Angry	0/35	0/35	25/35	10/35
Happy	1/27	4/27	8/27	14/27

Table 11.10 Confusion matrix after Stage 2 for EMO-DB database [66]

	Neutral	Sad	Angry	Happy
Neutral	74/79	4/79	0/79	1/79
Sad	27/62	33/62	0/62	2/62
Angry	2/127	0/127	114/127	11/127
Happy	7/71	3/71	27/71	34/71

The C_q of a glottal pulse is the ratio of the closed phase duration to the duration of the total glottal pulse, and is denoted by γ in percentage. The open and closed phases of a glottal cycle are illustrated in Fig. 11.2 through the EGG and the derivative of EGG (dEGG) signals. The amplitude of the EGG (current flow) signal is larger during the close phase region due to low impedance and lower during the open phase region due to high impedance across vocal folds. The locations of the GCI are associated with positive peaks, and the locations of the glottal opening instant are associated with negative peaks in the dEGG signal. The average (A_γ) values of γ for neutral, happiness and anger emotions, for five speakers of IIIT-H Telugu and EMO-DB databases are given in Tables 11.11 and 11.12, respectively. From the values of A_γ, it can be observed that the C_q has increasing trend in happiness and anger emotions for a given speaker, with anger possessing relatively higher value.

The ratio of high (800–5000 Hz) to low (0–400 Hz) frequency band energy is denoted as β. The β values are computed from the short time Fourier transform. In [86], it was reported that, increase in C_q increases the value of β. The average (A_β) values of β and the average (A_{SoE}) values of SoE for neutral, happiness and anger emotions for 5 speakers of IIIT-H Telugu and EMO-DB databases are shown in Tables 11.11 and 11.12, respectively.

Tables 11.11 and 11.12 show that the values of A_β and A_{SoE} can be related to the value of C_q. Increase in C_q (γ in percentage) is observed with increase in β and decrease in SoE. There is increasing trend in the values of A_β and decreasing trend in the values of A_{SoE} for happiness and anger emotions w.r.t. neutral. The values of A_β are relatively higher, and the values of A_{SoE} are relatively lower in case of anger when compared to the case of happiness. Although these features carry discriminative property, the dynamic ranges of their values are speaker-specific.

A speaker-specific anger versus happiness classification is implemented. A sample 2-D feature distributions (β versus SoE) of happy and angry utterances is given in

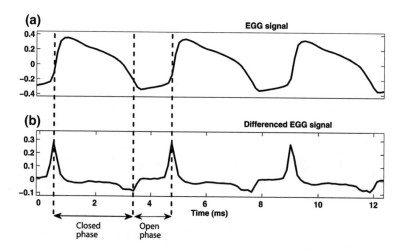

Fig. 11.2 Illustration of open and closed phase durations. **a** EGG signal. **b** dEGG signal [86]

Table 11.11 Average (A_γ, A_β and A_{SoE}) values of γ, β and SoE of neutral, happy and angry utterances for 5 speakers of IIIT-H Telugu database [43]

	Neutral			Happy			Angry		
	A_γ	A_β	A_{SoE}	A_γ	A_β	A_{SoE}	A_γ	A_β	A_{SoE}
Speaker 1	42.21	8.63	43.53	48.38	12.89	21.63	53.12	13.32	13.47
Speaker 2	39.86	9.86	28.93	42.19	13.26	13.91	44.31	14.26	7.76
Speaker 3	41.46	7.34	37.21	46.32	13.90	19.12	51.74	13.95	11.49
Speaker 4	44.55	7.40	18.97	44.19	13.62	7.76	46.14	13.81	5.67
Speaker 5	41.87	8.78	16.45	47.43	13.56	4.98	47.23	13.96	4.04
Mean value	41.99	8.40	29.02	45.70	13.45	13.48	48.51	13.86	8.49

Table 11.12 Average (A_γ, A_β and A_{SoE}) values of γ, β and SoE of neutral, happy and angry utterances for 5 speakers of EMO-DB database [43]

	Neutral			Happy			Angry		
	A_γ	A_β	A_{SoE}	A_γ	A_β	A_{SoE}	A_γ	A_β	A_{SoE}
Speaker 1	42.97	7.23	23.42	48.91	10.44	12.43	50.63	12.45	9.43
Speaker 2	44.14	6.32	19.34	47.89	11.02	11.34	51.01	11.53	8.32
Speaker 3	39.26	8.21	17.44	46.21	9.87	12.23	48.54	13.47	10.21
Speaker 4	44.04	5.56	11.21	48.24	8.32	3.96	49.66	9.61	3.32
Speaker 5	44.83	8.30	12.56	48.56	11.29	6.23	49.15	13.81	4.92
Mean value	43.05	7.12	16.79	47.96	10.19	9.24	49.80	12.17	7.24

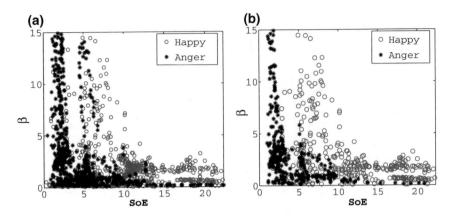

Fig. 11.3 Feature distributions (β versus SoE) of happy and angry emotion utterances. **a** Entire utterance. **b** High F_0 regions of the utterance [43]

Fig. 11.3a. As these emotions are produced suddenly in response to a particular stimuli, they are not sustainable. There may be neutral regions in these utterances [88]. To reduce the effect of neutral regions, the regions corresponding to high F_0 (above the average F_0) are considered. This can be justified as the speech in happiness and anger states is produced with increased pitch levels. The 2-D feature distributions of high F_0 regions of the happy and angry utterances are given in Fig. 11.3b. From Fig. 11.3a, b, it is evident that there is better discrimination between happiness and anger in the high F_0 regions.

Given a set of happy and angry utterances, the 2-D feature distributions (β versus SoE) of all utterances are computed. The feature distribution with low mean SoE and high mean β is considered as reference for anger emotion, and the feature distribution with high mean SoE and low mean β is considered as reference for happiness emotion. The remaining feature distributions are classified by computing the KL distance with the reference feature distributions. The lower the KL distance, the closer are the test distributions towards the reference. The classification accuracy of approximately 85 % is observed for both the databases.

In general, the positive emotions possesses more rhythmic behavior than the negative emotions [88]. To exploit this characteristic, the variances of SoE values are examined. The variance of SoE corresponding to high F_0 regions is compared with the variance of the entire utterance. A parameter called relative SoE variance is defined. This is given by

$$S_v = \frac{V_h - V_t}{\left(\frac{\mu_h + \mu_t}{2}\right)}, \tag{11.4}$$

where V_h and V_t are the SoE variances in the selected high F_0 regions and for the entire utterance, respectively, and μ_h and μ_t are the corresponding means. The average value of S_v of happy and angry utterances of IIIT-Telugu and EMO-DB databases is given in Table 11.13. It is observed that the S_v value for happy utterances is mostly

Table 11.13 Average value of S_v of happy and angry utterances of IIIT-H Telugu and German EMO-DB databases [43]

	IIIT-H Telugu		German EMO-DB	
	Happy	Angry	Happy	Angry
Average S_v	0.42	−0.51	−0.21	−0.58

positive in the case of IIIT-H Telugu database. The S_v value for angry utterances is observed to be mostly negative in both the databases. Sample histograms of happy and angry utterances are given in Fig. 11.4. Following this observation, a classification approach is proposed. The percentage of accuracy of 75 and 68 % are observed for IIIT-H Telugu and EMO-DB databases, respectively.

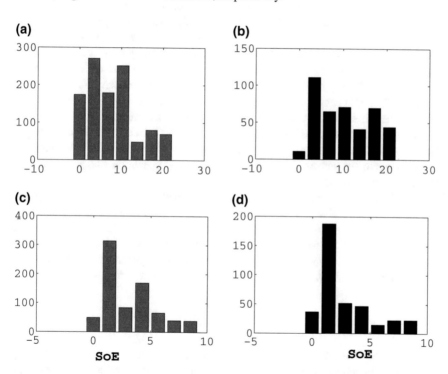

Fig. 11.4 Histograms of *SoE* values of happy and angry utterances. **a** Entire happy utterance. **b** High F_0 regions of the happy utterance. **c** Entire angry utterance. **d** High F_0 regions of the angry utterance [43]

11.4.6 Discrimination Between High Arousal and Falsetto Voices

High arousal speech is produced by a speaker in situations like emotionally charged states, communication over a long distance and in noisy environments. High arousal speech is often produced with increased levels of voice intensity. In natural communication, speakers tend to increase their pitch in high arousal speech. Increase in pitch can also be intentional as in falsetto register. This study is to discriminate high arousal and falsetto voices, as either of these cases there is increase in the average F_0 w.r.t. neutral.

The vocal folds configuration is significantly different in modal and falsetto registers [124, 130]. It is observed that the C_q of the vocal fold vibration is lower for segments of falsetto voice when compared to that of modal voice.

The average (A_γ) values of γ and the average (A_{F_0}) values of F_0 of neutral, happy, angry, shout and falsetto utterances for a set of five speakers from IIIT-H Telugu database are given in Table 11.14. On observation of A_{F_0} values, it is evident that falsetto modes are produced with distinct pitch variations. It is also clear that the F_0 range of low falsetto voice matches well with that of high arousal. The values of A_γ increases for happy, angry and shout speech w.r.t. neutral speech. For all the cases of falsetto, there is significant decrease in the A_γ values. It is interesting to note that the A_γ values are decreasing with the ascending degree of falsetto. The inconsistencies for different falsetto cases may be because of the lack of control in the voice intensity while producing speech at higher pitch levels by these speakers, who are not trained to produce these voices.

The glottal vibration characteristics also have an effect on the vocal tract system response. The effective length of the vocal tract changes in the open and closed regions of the vocal folds. Therefore, useful information of the excitation source characteristics can be obtained from speech analysis with high spectro-temporal resolution. A recently proposed zero time windowing (ZTW) method [135] is useful for this analysis. A heavily tapering window in the temporal domain, given by

$$h[n] = \frac{1}{8sin(\frac{\omega n}{N})^4}, \tag{11.5}$$

is used at each sampling instant. The Hilbert envelope of the numerator of group delay function (HNGD) of the windowed segment is computed at every sampling instant.

The β feature, which is the ratio of energies in the high (800–5000 Hz) and low (0–400 Hz) frequency bands of the HNGD spectra is computed. The β contour for a speech segment is shown in Fig. 11.5. The sharp peaks in the β contour occur mainly at GCIs. The β contour has lower values during the open phase.

The highest and the lowest values of the β contour within a glottal cycle are denoted as β_c and β_o, respectively. The average $(A_{\beta_c}$ and $A_{\beta_o})$ values of β_c and β_o, respectively, of neutral, happy, angry, shout and falsetto utterances are given in

Table 11.14 Average (A_γ and A_{F_0} in Hz) values of γ and F_0 of neutral, happy, angry, shout and falsetto utterances for 5 speakers

	Neutral		Happy		Angry		Shout		Low falsetto		Mid falsetto		High falsetto	
	A_γ	A_{F_0}	A_γ	A_{F_0}	A_γ	A_{F_0}	A_γ	A_{F_0}	A_γ	A_{F_0}	A_γ	A_{F_0}	A_γ	A_{F_0}
Speaker 1	42.21	135	48.38	161	53.12	197	56.21	257	33.42	213	31.23	362	35.13	428
Speaker 2	39.86	248	42.19	326	44.31	311	48.92	331	37.87	328	39.13	413	42.14	527
Speaker 3	41.46	142	46.32	198	51.74	241	50.46	238	32.19	321	33.12	376	37.98	371
Speaker 4	44.55	221	44.19	239	46.14	318	53.16	326	37.32	367	36.21	432	41.19	463
Speaker 5	41.87	162	47.43	245	47.23	287	46.37	328	34.42	340	34.56	444	39.17	522
Mean value	41.99	182	45.70	234	48.51	271	51.02	296	35.04	314	34.85	405	39.12	462

Table 11.15 Average (A_{β_c} and A_{β_o}) values of β_c and β_o of neutral, happy, angry, shout and falsetto utterances for 5 speakers

	Neutral		Happy		Angry		Shout		Low falsetto		Mid falsetto		High falsetto	
	A_{β_c}	A_{β_o}	A_{β_c}	A_{β_o}	A_{β_c}	A_{β_o}	A_{β_c}	A_{β_o}	A_{β_c}	A_{β_o}	A_{β_c}	A_{β_o}	A_{β_c}	A_{β_o}
Speaker 1	0.21	0.03	0.37	0.05	0.41	0.11	0.43	0.17	0.22	0.07	0.33	0.16	0.39	0.21
Speaker 2	0.51	0.10	0.55	0.08	0.82	0.24	1.11	0.29	0.48	0.27	1.36	0.64	1.9	0.96
Speaker 3	0.27	0.04	0.33	0.04	0.39	0.09	0.45	0.19	0.25	0.08	0.26	0.1	0.48	0.31
Speaker 4	0.49	0.09	0.63	0.04	1.21	0.31	1.40	0.36	0.47	0.27	1.36	0.63	1.51	0.69
Speaker 5	0.23	0.03	0.23	0.02	0.29	0.05	0.39	0.14	0.25	0.07	0.41	0.21	0.53	0.26
Mean value	0.34	0.06	0.42	0.05	0.62	0.16	0.76	0.23	0.33	0.15	0.74	0.35	0.96	0.49

Fig. 11.5 a Segment of a speech signal. **b** Differenced EGG signal. **c** β contour

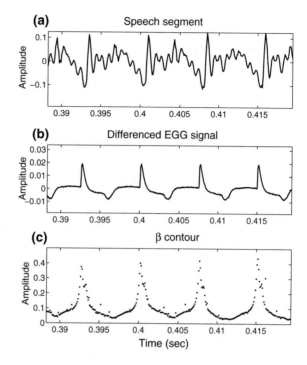

Table 11.15. From Table 11.15, it is evident that there is significant increase in A_{β_c} for angry and shout speech. A similar, but not significant trend can be observed in the case of happy speech. From these results, we can infer that increase in C_q increases A_{β_c}. In the case of low falsetto voice, A_{β_c} is in similar range as for the neutral. But, the results are not consistent for mid and high falsetto voices.

11.5 Research Challenges in Emotional Speech Analysis

The major issues in dealing with emotional speech analysis are the description of emotion, data collection and feature representation. The production and perception of emotions in speech by human beings is a complex phenomenon, which is not well understood. A basic and widely accepted statement is that emotions are underlying states of a speaker, and emotion expressions are the communicative value of the underlying states [22]. In natural communication, speakers may sometimes try to hide the underlying emotional state. It is very difficult to collect spontaneous emotion data, and also it is difficult to annotate the collected data. This statement is supported by the fact that many analysis studies in the literature use full-blown emotional data. Full-blown emotional speech corresponds to intense expressions of underlying emotional states [104].

The features used for analysis are broadly categorized as prosody, voice quality and spectral features. These features carry emotion correlates, but they are observed to be speaker and sound-specific. To derive an emotion-specific feature representation, the speech production knowledge of emotions is required. In general, speech in different emotion states is produced with distinct changes in speech production mechanism. However, it is difficult to describe these emotion-specific deviations in the production mechanism. This makes emotional speech analysis a difficult task.

The state-of-the-art approaches for emotion recognition are adopted from the developments in applications like speech recognition and speaker recognition. In these approaches, the utterances are labeled with the discrete emotion or primitive categories. Pattern recognition models like GMMs, SVMs, ANNs and HMMs are trained on the features extracted. As most of the feature representations are speaker and sound-specific, these pattern recognition models require a phonetically/phonemically balanced data, covering several speakers. In practical sense, it is difficult to collect such a database. Given the limitations of existing features and data collection issues, the ideal scenario is to identify underlying emotion in speech by a feature representation.

The experiments conducted in our research show that the excitation source features carry significant amount of emotion related information, and these features are also observed to be speaker-specific. It is indeed a challenge to determine the excitation source features that are only emotion-specific.

Although it is difficult to define production characteristics that are specific to an emotion, there are clues when emotions are viewed in the 3 dimensions/primitives (arousal, valence and dominance). There is increase in F_0 in the case of high arousal speech, and decrease in F_0 in the case of low arousal speech. But there may be cases where speech can be produced with deliberate increase/decrease in F_0. Discrimination of speech with natural increase/decrease in F_0 and deliberate increase/decrease in F_0 is an important issue in emotion studies. In the case of valence dimension, in [88], it was reported that rhythm and valence were consistently related. The positive feelings possess regular rhythm than negative feelings. From our studies (reported in Sects. 11.4.6 and 11.4.5), it can be said that feature representation of primitives of emotion might help in representing emotional speech effectively. From the current studies, it appears that emotion recognition by a machine appears to be an elusive goal.

References

1. Airas M, Alku P (2004) Emotions in short vowel segments: effects of the glottal flow as reflected by the normalized amplitude quotient. In: Affective dialogue systems. Springer, pp 13–24
2. Airas M, Pulakka H, Bäckström T, Alku P (2005) A toolkit for voice inverse filtering and parametrization. In: INTERSPEECH. Lisbon, Portugal, pp 2145–2148
3. Alku P (2011) Glottal inverse filtering analysis of human voice production a review of estimation and parameterization methods of the glottal excitation and their applications. Sadhana 36(5):623–650

4. Alku P, Vilkman E (1996) A comparison of glottal voice source quantification parameters in breathy, normal and pressed phonation of female and male speakers. Folia Phoniatrica et Logopaedica 48:240–254

5. Amer MR, Siddiquie B, Richey C, Divakaran A (2014) Emotion recognition in speech using deep networks. In: ICASSP. Florence, Italy, pp 3752–3756

6. Amir N, Kerret O, Karlinski D (2001) Classifying emotions in speech: a comparison of methods. In: INTERSPEECH. Aalborg, Denmark, pp 127–130

7. Ang j, Dhillon R, Krupski A, Shriberg E, Stolcke A (2002) Prosody-based automatic detection of annoyance and frustration in human-computer dialog. In: INTERSPEECH. Denver, Colorado, USA

8. Arias JP, Busso C, Yoma NB (2013) Energy and F0 contour modeling with functional data analysis for emotional speech detection. In: INTERSPEECH. Lyon, France, pp 2871–2875

9. Arias JP, Busso C, Yoma NB (2014) Shape-based modeling of the fundamental frequency contour for emotion detection in speech. Comput Speech Lang 28(1):278–294

10. Atassi H, Esposito A (2008) A speaker independent approach to the classification of emotional vocal expressions. In: IEEE international conference on tools with artificial intelligence (ICTAI'08), vol 2. Dayton, Ohio, USA, pp 147–152

11. Atassi H, Riviello M, Smékal Z, Hussain A, Esposito A (2010) Emotional vocal expressions recognition using the COST 2102 Italian database of emotional speech. In: Esposito A, Campbell N, Vogel C, Hussain A, Nijholt A (eds) Development of multimodal interfaces: active listening and synchrony. Lecture notes in computer science, vol 5967. Springer, Berlin, pp 255–267

12. Bachorowski J (1999) Vocal expression and perception of emotion. Curr Dir Psychol Sci 8(2):53–57

13. Banse R, Scherer KR (1996) Acoustic profiles in vocal emotion expression. J Personal Soc Psychol 70(3):614–636

14. Batliner A, Schuller B, Seppi D, Steidl S, Devillers L, Vidrascu L, Vogt T, Aharonson V, Amir N (2011) The automatic recognition of emotions in speech. In: Petta P, Pelachaud C, Cowie R (eds) Emotion-oriented systems. Springer, pp 71–99

15. Bezooijen RAMG, Otto SA, Heenan TA (1983) Recognition of vocal expressions of emotion: a three-nation study to identify universal characteristics. J Cross-Cult Psychol 14:387–406

16. Boersma P, Heuven VV (2001) Speak and unSpeak with PRAAT. Glot Int 5(9/10):341–347

17. Burkhardt F, Paeschke A, Rolfes M, Sendlmeier WF, Weiss B (2005) A database of German emotional speech. In: INTERSPEECH. Lisbon, Portugal, pp 1517–1520

18. Busso C, Bulut M, Lee C, Kazemzadeh A, Mower E, Kim S, Chang JN, Lee S, Narayanan S (2008) IEMOCAP: interactive emotional dyadic motion capture database. Lang Res Eval 42(4):335–359

19. Chastagnol C, Devillers L (2011) Analysis of anger across several agent-customer interactions in French call centers. In: ICASSP. Prague, Czech Republic, pp 4960–4963

20. Childers DG, Lee CK (1991) Vocal quality factors: analysis, synthesis, and perception. J Acoust Soc Am 90(5):2394–2410

21. Cowie R, Cornelius RR (2003) Describing the emotional states that are expressed in speech. Speech Commun 40(1–2):5–32

22. Darwin C (1872) The expression of emotion in man and animals. reprinted by University of Chicago Press, Murray, London, UK (1975)

23. Davitz JR (1964) Personality, perceptual, and cognitive correlates of emotional sensitivity. In: Davitz JR (ed) The communication of emotional meaning. McGraw-Hill, New York

24. Dellaert F, Polzin T, Waibel A (1996) Recognizing emotion in speech. In: international conference on spoken language processing (ICSLP). Philadelphia, USA, pp 1970–1973

25. Devillers L, Vidrascu L (2006) Real-life emotions detection with lexical and paralinguistic cues on human-human call center dialogs. In: INTERSPEECH. Pittsburgh, PA, USA, pp 801–804

26. Douglas-Cowie E, Campbell N, Cowie R, Roach P (2003) Emotional speech: towards a new generation of databases. Speech Commun 40(1–2):33–60

27. Ekman P (1992) An argument for basic emotions. Cognit Emot 6:169–200
28. Engberg IS, Hansen AV, Andersen O, Dalsgaard P (1997) Design, recording and verification of a Danish emotional speech database. In: EUROSPEECH. Rhodes, Greece, pp 1695–1698
29. Erden M, Arslan LM (2011) Automatic detection of anger in human-human call center dialogs. In: INTERSPEECH. Florence, Italy, pp 81–84
30. Erickson D, Yoshida K, Menezes C, Fujino A, Mochida T, Shibuya Y (2006) Exploratory study of some acoustic and articulatory characteristics of sad speech. Phonetica 63:1–5
31. Erro D, Navas E, Hernáez I, Saratxaga I (2010) Emotion conversion based on prosodic unit selection. IEEE Trans Audio Speech Lang Process 18(5):974–983
32. Espinosa HP, Garcia JO, Pineda LV (2010) Features selection for primitives estimation on emotional speech. In: ICASSP. Florence, Italy, pp 5138–5141
33. Eyben F, Wollmer M, Schuller B (2009) OpenEarIntroducing the Munich open-source emotion and affect recognition toolkit. In: International conference on affective computing and intelligent interaction and workshops (ACII). Amsterdam, Netherlands, pp 1–6
34. Eyben F, Batliner A, Schuller B, Seppi D, Steidl S (2010) Cross-corpus classification of realistic emotions—some pilot experiments. In: International workshop on EMOTION (satellite of LREC): corpora for research on emotion and affect. Valletta, Malta, pp 77–82
35. Eyben F, Wöllmer M, Schuller B (2010) OpenSMILE: The Munich versatile and fast open-source audio feature extractor. In: International conference on multimedia. Firenze, Italy, pp 1459–1462
36. Fairbanks G, Hoaglin LW (1941) An experimental study of the durational characteristics of the voice during the expression of emotion. Speech Monogr 8:85–91
37. Fairbanks G, Pronovost W (1939) An experimental study of the pitch characteristics of the voice during the expression of emotion. Speech Monogr 6:87–104
38. Fant G, Lin Q, Gobl C (1985) Notes on glottal flow interaction. Speech Transm Lab Q Progress Status Rep, KTH 26:21–25
39. Fernandez R, Picard R (2011) Recognizing affect from speech prosody using hierarchical graphical models. Speech Commun 53(9–10):1088–1103
40. Fonagy I, Magdics K (1963) Emotional patterns in intonation and music. Kommunikationforsch 16:293–326
41. Gangamohan P, Mittal VK, Yegnanarayana B (2012) A flexible analysis and synthesis tool (FAST) for studying the characteristic features of emotion in speech. In: IEEE international conference on consumer communications and networking conference. Las Vegas, USA pp 266–270
42. Gangamohan P, Sudarsana RK, Yegnanarayana B (2013) Analysis of emotional speech at subsegmental level. In: INTERSPEECH. Lyon, France, pp 1916–1920
43. Gangamohan P, Sudarsana RK, Suryakanth VG, Yegnanarayana B (2014) Excitation source features for discrimination of anger and happy emotions. In: INTERSPEECH. Singapore, pp 1253–1257
44. Gnjatovic M, Rösner D (2010) Inducing genuine emotions in simulated speech-based human-machine interaction: the nimitek corpus. IEEE Trans Affect Comput 1(2):132–144
45. Gobl C (1988) Voice source dynamics in connected speech. Speech Trans Lab Q Progress Status Rep, KTH 1:123–159
46. Gobl C (1989) A preliminary study of acoustic voice quality correlates. Speech Trans Lab Q Progress Status Rep, KTH 4:9–21
47. Gobl C, Chasaide AN (1992) Acoustic characteristics of voice quality. Speech Commun 11(4):481–490
48. Gobl C, Chasaide AN (2003) The role of voice quality in communicating emotion, mood and attitude. Speech Commun 40(1–2):189–212
49. Grichkovtsova I, Morel M, Lacheret A (2012) The role of voice quality and prosodic contour in affective speech perception. Speech Commun 54(3):414–429
50. Grimm M, Kroschel K, Mower E, Narayanan S (2007) Primitives-based evaluation and estimation of emotions in speech. Speech Commun 49(10–11):787–800

51. Grimm M, Kroschel K, Narayanan S (2008) The Vera am Mittag German audio-visual emotional speech database. In: International conference on multimedia and expo. Hannover, Germany, pp 865–868
52. Guruprasad S, Yegnanarayana B (2009) Perceived loudness of speech based on the characteristics of glottal excitation source. J Acoust Soc Am 126(4):2061–2071
53. Hansen JH, Womack BD (1996) Feature analysis and neural network-based classification of speech under stress. IEEE Trans Speech Audio Process 4(4):307–313
54. Hanson HM (1997) Glottal characteristics of female speakers: acoustic correlates. J Acoust Soc Am 101(1):466–481
55. Hassan A, Damper RI (2010) Multi-class and hierarchical SVMs for emotion recognition. In: INTERSPEECH. Chiba, Japan, pp 2354–2357
56. He L, Lech M, Allen N (2010) On the importance of glottal flow spectral energy for the recognition of emotions in speech. In: INTERSPEECH. Chiba, Japan, pp 2346–2349
57. Hershey JR, Olsen PA (2007) Approximating the Kullback Leibler divergence between Gaussian mixture models. In: ICASSP, vol 4. Montreal, Quebec, Canada, pp 317–320
58. Huber R, Batliner A, Buckow J, Nöth E, Warnke V, Niemann H (2000) Recognition of emotion in a realistic dialogue scenario. In: Proceedings of international conference on spoken language processing. Beijing, China, pp 665–668
59. Hübner D, Vlasenko B, Grosser T, Wendemuth A (2010) Determining optimal features for emotion recognition from speech by applying an evolutionary algorithm. In: INTERSPEECH. Chiba, Japan, pp 2358–2361
60. Izard CE (1977) Human emotions. Plenum Press, New York
61. Jeon JH, Xia R, Liu Y (2011) Sentence level emotion recognition based on decisions from subsentence segments. In: ICASSP. Lyon, France, pp 4940–4943
62. Jeon JH, Le D, Xia R, Liu Y (2013) A preliminary study of cross-lingual emotion recognition from speech: automatic classification versus human perception. In: INTERSPEECH. Prague, Czech Republic, pp 2837–2840
63. Joachims T (1998) Text categorization with support vector machines: Learning with many relevant features. In: European conference on machine learning. London, UK, pp 137–142
64. Kadiri SR, Gangamohan P, Mittal VK, Yegnanarayana B (2014) Naturalistic audio-visual emotion database. In: International conference on natural language processing. Goa, India, pp 127–134
65. Kadiri SR, Gangamohan P, Yegnanarayana B (2014) Discriminating neutral and emotional speech using neural networks. In: Interenational conference on natural language processing. Goa, India, pp 119–126
66. Kadiri SR, Gangamohan P, Gangashetty SV, Yegnanarayana B (2015) Analysis of excitation source features of speech for emotion recognition. In: INTERSPEECH. Dresden, Germany, pp 1032–1036
67. Keller E (2005) The analysis of voice quality in speech processing. In: Gèrard C, Anna E, Marcos F, Maria M (eds) Lecture notes in computer science. Springer, pp 54–73
68. Kim W, Hansen JHL (2010) Angry emotion detection from real-life conversational speech by leveraging content structure. In: ICASSP. Dallas, Texas, USA, pp 5166–5169
69. Kim J, Lee S, Narayanan S (2010) An exploratory study of manifolds of emotional speech. In: ICASSP. Dallas, Texas, USA, pp 5142–5145
70. Kim J, Park J, Oh Y (2011) On-line speaker adaptation based emotion recognition using incremental emotional information. In: ICASSP. Prague, Czech Republic, pp 4948–4951
71. Klasmeyer G, Sendlmeier WF (2000) Voice and emotional states. In: Voice quality measurement. Springer, Berlin, Germany, pp 339–358
72. Klatt DH (1980) Software for a cascade/parallel formant synthesizer. J Acoust Soc Am 67(3):971–995
73. Koolagudi SG, Sreenivasa Rao K (2012) Emotion recognition from speech: a review. Int J Speech Technol 15(2):99–117
74. Koolagudi SG, Maity S, Vuppala AK, Chakrabarti S, Sreenivasa Rao K (2009) IITKGP-SESC: speech database for emotion analysis. In: Communications in computer and information science, pp 485–492

75. Laver John DM (1968) Voice quality and indexical information. Int J Lang Commun Disord 3(1):43–54
76. Lee C, Mower E, Busso C, Lee S, Narayanan S (2011) Emotion recognition using a hierarchical binary decision tree approach. Speech Commun 53(9–10):1162–1171
77. Lee CM, Narayanan S (2005) Toward detecting emotions in spoken dialogs. IEEE Trans Speech Audio Process 13(2):293–303
78. Lee CM, Yildirim S, Bulut M, Kazemzadeh A, Busso C, Deng Z, Lee S, Narayanan S (2004) Emotion recognition based on phoneme classes. In: INTERSPEECH. JejuIsland, Korea, pp 205–211
79. Lieberman P, Michaels SB (1962) Some aspects of fundamental frequency and envelope amplitude as related to the emotional content of speech. J Acoust Soc Am 34(7):922–927
80. Lin J, Wu C, Wei W (2013) Emotion recognition of conversational affective speech using temporal course modeling. In: INTERSPEECH. Lyon, France, pp 1336–1340
81. Luengo I, Navas E, Hernáez I, Sánchez J (2005) Automatic emotion recognition using prosodic parameters. In: INTERSPEECH. Lisbon, Portugal, pp 493–496
82. Lugger M, Yang B (2007) The relevance of voice quality features in speaker independent emotion recognition. In: ICASSP, vol 4. Honolulu, Hawaii, USA, pp 17–20
83. Makhoul J (1975) Linear prediction: a tutorial review. Proc IEEE 63:561–580
84. Mansoorizadeh M, Charkari NM (2007) Speech emotion recognition: comparison of speech segmentation approaches. In: Proceedings of IKT, Mashad, Iran
85. McGilloway S, Cowie R, Douglas-Cowie E, Gielen S, Westerdijk M, Stroeve S (2000) Approaching automatic recognition of emotion from voice: a rough benchmark. In: ISCA tutorial and research workshop (ITRW) on speech and emotion. Newcastle, Northern Ireland, UK
86. Mittal VK, Yegnanarayana B (2013) Effect of glottal dynamics in the production of shouted speech. J Acoust Soc Am 133(5):3050–3061
87. Morrison D, Wang R, De Silva LC (2007) Ensemble methods for spoken emotion recognition in call-centres. Speech Commun 49(2):98–112
88. Murray IR, Arnott JL (1993) Toward the simulation of emotion in synthetic speech: a review of the literature on human vocal emotion. J Acoust Soc Am 93(2):1097–1108
89. Murty KSR, Yegnanarayana B (2008) Epoch extraction from speech signals. IEEE Trans Audio Speech Lang Process 16(8):1602–1613
90. Nogueiras A, Moreno A, Bonafonte A, Mariño JB (2001) Speech emotion recognition using hidden Markov models. In: EUROSPEECH. Aalborg, Denmark, pp 2679–2682
91. Nwe TL, Foo SW, De Silva LC (2003) Speech emotion recognition using hidden Markov models. Speech Commun 41(4):603–623
92. Oatley K (1989) The importance of being emotional. New Sci 123:33–36
93. Pereira C (2000) Dimensions of emotional meaning in speech. In: ISCA tutorial and research workshop (ITRW) on speech and emotion. Northern Ireland, UK
94. Polzehl T, Sundaram S, Ketabdar H, Wagner M, Metze F (2009) Emotion classification in children's speech using fusion of acoustic and linguistic features. In: INTERSPEECH. Brighton, UK, pp 340–343
95. Prasanna SRM, Govind D (2010) Analysis of excitation source information in emotional speech. In: INTERSPEECH. Chiba, Japan, pp 781–784
96. Rothenberg M (1973) A new inverse-filtering technique for deriving the glottal air flow waveform during voicing. J Acoust Soc Am 53(6):1632–1645
97. Rozgic V, Ananthakrishnan S, Saleem S, Kumar R, Vembu AN, Prasad R (2012) Emotion recognition using acoustic and lexical features. In: INTERSPEECH. Portland, USA
98. Scherer KR (1981) Speech and emotional states. In: Darby JK (ed) Speech evaluation in psychiatry. Grune and Stratton, New York
99. Scherer KR (1984) On the nature and function of emotion: a component process approach. In: Scherer KR, Ekman P (eds) Approaches to emotion. Lawrence Elbraum, Hillsdale
100. Scherer KR (2003) Vocal communication of emotion: a review of research paradigms. Speech Commun 40(1–2):227–256

101. Scholsberg H (1941) A scale for the judgment of facial expressions. J Exp Psychol 29(6):497–510
102. Schlosberg H (1954) Three dimensions of emotion. J Psychol Rev 61(2):81–88
103. Schröder M (2001) Emotional speech synthesis-a review. In: INTERSPEECH. Aalborg,Denmark, pp 561–564
104. Schröder M (2004) Speech and emotion research: an overview of research frameworks and a dimensional approach to emotional speech synthesis. PhD thesis, Saarland University
105. Schröder M, Cowie R, Douglas-Cowie E, Westerdijk M, Gielen SC (2001) Acoustic correlates of emotion dimensions in view of speech synthesis. In: INTERSPEECH. Aalborg, Denmark, pp 87–90
106. Schuller B (2011) Recognizing affect from linguistic information in 3D continuous space. IEEE Trans Affect Comput 2(4):192–205
107. Schuller B, Rigoll G (2006) Timing levels in segment-based speech emotion recognition. In: INTERSPEECH. Pittsburgh, Pennsylvania, pp 17–21
108. Schuller B, Rigoll G, Lang M (2004) Speech emotion recognition combining acoustic features and linguistic information in a hybrid support vector machine-belief network architecture. In: ICASSP vol 1. Montreal, Quebec, Canada, pp 577–580
109. Schuller B, Müller R, Lang M, Rigoll G (2005) Speaker independent emotion recognition by early fusion of acoustic and linguistic features within ensembles. In: INTERSPEECH. Lisbon, Portugal, pp 805–808
110. Schuller B, Villar RJ, Rigoll G, Lang MK (2005) Meta-classifiers in acoustic and linguistic feature fusion-based affect recognition. In: ICASSP. Philadelphia, Pennsylvania, USA, pp 325–328
111. Schuller B, Batliner A, Steidl S, Seppi D (2009) Emotion recognition from speech: putting ASR in the loop. In: ICASSP. Taipei, Taiwan, pp 4585–4588
112. Schuller B, Batliner A, Steidl S, Seppi D (2011) Recognising realistic emotions and affect in speech: state of the art and lessons learnt from the first challenge. Speech Commun 53(9–10):1062–1087
113. Schuller B, Vlasenko B, Eyben F, Wollmer M, Stuhlsatz A, Wendemuth A, Rigoll G (2010) Cross-corpus acoustic emotion recognition: variances and strategies. IEEE Trans Affect Comput 1(2):119–131
114. Shami M, Verhelst W (2007) An evaluation of the robustness of existing supervised machine learning approaches to the classification of emotions in speech. Speech Commun 49(3):201–212
115. Shaver P, Schwartz J, kirson D, O'Connor C (1987) Emotion, knowledge: further exploration of a prototype approach. J Personal Soc Psychol 52:1061–1086
116. Sneddon I, McRorie M, McKeown G, Hanratty J (2012) The Belfast induced natural emotion database. IEEE Trans Affect Comput 3(1):32–41
117. Steidl S (2009) Automatic classification of emotion related user states in spontaneous children's speech. PhD thesis, Universität Erlangen-Nürnberg, Germany
118. Steidl S, Batliner A, Seppi D, Schuller B (2010) On the impact of children's emotional speech on acoustic and language models. EURASIP J Audio, Speech, and Music Processing
119. Stein N, Oatley K (1992) Basic emotions: theory and measurement. Cognit Emot 6:161–168
120. Sun R, Moore II E (2012) A preliminary study on cross-databases emotion recognition using the glottal features in speech. In: INTERSPEECH. Portland, USA, pp 1628–1631
121. Sun R, Moore II E, Torres JF (2009) Investigating glottal parameters for differentiating emotional categories with similar prosodics. In: ICASSP. Taipei, Taiwan, pp 4509–4512
122. Sundberg J, Patel S, Bjorkner E, Scherer KR (2011) Interdependencies among voice source parameters in emotional speech. IEEE Trans Affect Comput 2(3):162–174
123. Tahon M, Degottex G, Devillers L (2012) Usual voice quality features and glottal features for emotional valence detection. In: Speech Prosody. Shanghai, China, pp 693–696
124. Titze IR (1994) Principles of voice production. Prentice-Hall, Englewood Cliffs
125. Truong Khiet P, van Leeuwen David A, de Jong Franciska M G (2012) Speech-based recognition of self-reported and observed emotion in a dimensional space. Speech Commun 54(9):1049–1063

126. Ververidis D, Kotropoulos C (2003) A review of emotional speech databases. In: Proceedings of panhellenic conference on informatics (PCI). Thessaloniki, Greece, pp 560–574
127. Ververidis D, Kotropoulos C (2005) Emotional speech classification using Gaussian mixture models. In: International symposium on circuits and systems. Kobe, Japan, pp 2871–2874
128. Vlasenko B, Prylipko D, Philippou-Hübner D, Wendemuth A (2011) Vowels formants analysis allows straightforward detection of high arousal acted and spontaneous emotions. In: INTERSPEECH. Florence, Italy, pp 1577–1580
129. Vroomen J, Collier R, Mozziconacci S (1993) Duration and intonation in emotional speech. In: EUROSPEECH, vol 1. Berlin, Germany, pp 577–580
130. Švec Jan G, Schutte Harm K, Miller Donald G (1999) On pitch jumps between chest and falsetto registers in voice: data from living and excised human larynges. J Acoust Soc Am 106(3):1523–1531
131. Waaramaa T, Laukkanen AM, Airas M, Alku P (2010) Perception of emotional valences and activity levels from vowel segments of continuous speech. J Voice 24(1):30–38
132. Williams CE, Stevens KN (1969) On determining the emotional state of pilots during flight: an exploratory study. Aerosp Med 40:1369–1372
133. Williams CE, Stevens KN (1972) Emotions and speech: some acoustical correlates. J Acoust Soc Am 52(2):1238–1250
134. Wu S, Falk TH, Chan W (2011) Automatic speech emotion recognition using modulation spectral features. Speech Commun 53(5):768–785
135. Yegnanarayana B, Dhananjaya N (2013) Spectro-temporal analysis of speech signals using zero-time windowing and group delay function. Speech Commun 55(6):782–795
136. Yegnanarayana B, Murty KSR (2009) Event-based instantaneous fundamental frequency estimation from speech signals. IEEE Trans Audio Speech Lang Process 17(4):614–624
137. Yeh L, Chi T (2010) Spectro-temporal modulations for robust speech emotion recognition. In: INTERSPEECH. Chiba, Japan, pp 789–792
138. Zeng Z, Pantic M, Roisman GI, Huang TS (2009) A survey of affect recognition methods: Audio, visual, and spontaneous expressions. IEEE Trans Pattern Anal Mach Intell 31(1):39–58

Author Index

Printed in the United States
By Bookmasters